D0969320

Blowout in the Gulf

Blowout in the Gulf

The BP Oil Spill Disaster and the Future of Energy in America

William R. Freudenburg and Robert Gramling

The MIT Press
Cambridge, Massachusetts
London, England

For information about special quantity discounts, please email special_sales@mitpress.mit.edu

This book was set in Sabon by the MIT Press. Printed and bound in the United States of America.

Library of Congress Cataloging-in-Publication Data

Freudenburg, William R.
Blowout in the Gulf : the BP oil spill disaster and the future of energy in America / William R. Freudenburg and Robert Gramling.
p. cm.
Includes bibliographical references and index.
ISBN 978-0-262-01583-7 (hardcover : alk. paper)
1. BP Deepwater Horizon Explosion and Oil Spill, 2010. 2. Oil spills—Mexico, Gulf of. 3. Drilling platforms—Accidents—Mexico, Gulf of. 4. Offshore oil industry—Environmental aspects—Mexico, Gulf of. I. Gramling, Robert, 1943– II. Title.
GC1221.F74 2011

2010937510

10 9 8 7 6 5 4 3 2 1

Contents

Prologue: The Deep-Water Horror Zone ix

1 A Question for Our Time 1

2 The Macondo Mess 9

3 Stored Sunlight and Its Risks 21

4 Colonel of an Industry 63

5 Barons and Barrels 75

6 Off the Edge in All Directions 91

7 "Energy Independence" 113

8 To Know Us Is to Love Us? 129

9 Cleaning Up 153

10 Today and Tomorrow 171

Notes 191
References 203
Index 225

To Sarah, Max and Eileen

Prologue: The Deep-water Horror Zone

April 20, 2010, had been a pretty good day for the friends on the 26-foot craft, *Endorfin*. Fishing for blackfin tuna, they had caught their limit, and as night fell, they headed toward the *Deepwater Horizon*—a gigantic drilling rig that had been enjoying a pretty good day as well.

Just seven months earlier, the big rig had set an all-time record for deepwater drilling, completing a well nearly six miles deep. The day before, one of the platform's key contractors—Halliburton—had finished cementing the current well's final casing, a key step in the process of getting the platform ready to move to a new location. Topping things off, April 20 was the day when important corporate bigwigs had come on board, celebrating the fact that the *Deepwater Horizon* had just completed seven full years without a single lost-time accident—the first such rig ever to do so.[1]

As would befit its record-setting status, the *Deepwater Horizon* was a marvel of technology. In many ways, it was more of a ship than a drilling platform—two submarine-like hulls, floating below the surface, where waves had little effect, plus a deck up above the waves that provided living and working space for the crew. In other ways, though, it was more of a city than a ship—a complex of steel and machinery, served around the clock by a crew of 130, and with a deck as big as two football fields, floating side by side. Also like a city, the *Deepwater Horizon* was intended to stay in one spot, at least once it reached a drilling location, using global positioning technology so precise that the its drills could hit a specific spot on the ocean floor, just inches in diameter, but located nearly a mile below.

The earliest exploratory offshore drilling rigs had a much easier task of lining things up; they sat in one spot or stood on tall steel "legs" firmly attached to the bottom of the sea. As the drilling moved to ever-deeper locations, though, it became impossibly expensive to build rigs that could support themselves from the sea bottoms, thousands of feet below. Instead, oil companies shifted to new technologies—"semisubmersible" rigs or drill ships, floating on the surface rather than standing on the bottom. Early semisubmersibles were tethered in one spot by using a set of cables and anchors. Those cables continued to work well, even as water got deep enough to crush a Navy submarine, but in the spot where the *Deepwater Horizon* was drilling—an area known as Mississippi Canyon block 252—the water was almost a mile deep. A tethered drilling rig in that location would have required an almost prohibitively heavy, expensive, and complicated set of anchors, connected with cables that would have needed to be miles long. Instead, the drilling rig used a set of eight massive thrusters—each one capable of producing over 7,000 horsepower—in a complex choreography that kept the rig precisely aligned.

On the *Endorfin* on April 20 were Albert Andry III, a student in Marine Biology, and three of his high-school friends. Fishing and oil drilling had a long history of coexistence in the Gulf, and the friends intended to idle through the night at the massive drilling operation. When they first got to the rig, things looked particularly serene—the sea was as calm as the surface of a mirror—and they started catching bait for the next day's fishing. Just after 9:30 that night, though, things suddenly got anything but peaceful. Water came gushing down so fast that Andry thought the *Deepwatwer Horizon* crew was dumping its bilge water to keep from capsizing, and the friends' eyes started to burn. One of them who had experience working on rigs, Wes Bourg, knew that they needed to move fast, shouting to his friend to "Go, go, go, go, GOOOOO!" Andry gunned the throttle and headed for open water as fast as his boat could go. The *Endorfin* was about 100 yards away when the platform exploded into flames.[2]

By the narrowest of margins, the friends on board the *Endorfin* all survived. Above them, though, the crew members of the *Deepwater Horizon* were not so fortunate—and neither were the wildlife or the other human inhabitants of the Gulf region. Seventeen of the crew members suffered serious injuries, and eleven more were killed in the explosion. In just the first few weeks after the spill, several hundred sea turtles, all of them officially threatened or endangered, washed up dead. They were joined by hundreds of porpoises and other sea mammals, thousands of seabirds, and an unknowable number of fish, which would die from the spilled oil, from the dispersants that were used in an effort to break up that oil, or both. On shore, meanwhile, the millions of human inhabitants of the Gulf coast states, slowly starting to recover from the devastation of Hurricanes Katrina, Rita, and Ike, were about to be confronted by a new disaster.[3]

For some of the workers who managed to survive the initial explosion, the force was enough to knock them off their feet or to bury them under debris. Struggling through smoke, heat and darkness, most managed to reach the lifeboats that were being lowered to the surface of the Gulf, some 80 feet below, but some had to jump, hitting the surface of the water with a force of 20 Gs.

The Coast Guard was contacted almost immediately; the service ship *Joe Griffin*, equipped with water cannons that could pump out 10,000 gallons of water a minute, managed to fire up its engines and get underway in a quarter of the time usually required. Unfortunately, although the *Joe Griffin* was heading out toward the burning rig at full speed, that meant was the trip out to the rig would take more than nine and a half hours. The glow of the flames were visible from 35 miles away.

The effort to put out the flames was heroic, but futile. Thirty-six hours later, during the late morning hours of April 22, in a strange but spectacular commemoration, the charred remains of the *Deepwater Horizon* collapsed and sank to the bottom of the sea.[4]

It was the fortieth anniversary of Earth Day.

The initial assessments of the spill ranged from the argument by BP's CEO, Tony Hayward—namely that "The overall environmental impact of this will be very, very modest"—to the declaration by President Obama, and many others, that the spill will ultimately be seen as "the worst environmental disaster America has ever faced." At the moment, the long-term outcomes of that debate are no more clear than are the waters of the Gulf. Instead, based on our experience in dealing with other disasters, we can already offer the confident prediction that variations on these same arguments will continue to be made for decades to come, providing a steady income to lawyers yet unborn.

Even at this early stage, however, it is possible to start bringing much greater clarity to our thinking. The key to doing so is by focusing on some of the larger lessons that are available to be learned from this and other disasters. That is particularly true with the lesson that will be the major focus of this book, which applies not just to BP, but also more broadly: Both literally and figuratively, and both in the Gulf of Mexico and elsewhere, we have been getting into increasingly dangerous waters, doing so without being sufficiently vigilant about the implications of our actions. Perhaps the logical place to start, then, is by asking why the crew of the *Deepwater Horizon* would have been working in such a dangerous spot in the first place.

1 A Question for Our Time

When future historians look back on the first decade of the twenty-first century, they are likely to focus much of their attention on the dramatic images provided by the U.S. invasions of Afghanistan and Iraq. Millions of Americans watched as the tanks rolled into Baghdad, where a small crowd of happy Iraqis cheered as the tanks pulled down the hollow statue of Saddam Hussein. Soon after that, unfortunately, Americans also learned about less-happy Iraqis who were exploding homemade bombs and shooting rocket-propelled grenades at some of those very same tanks.

Far less visible or dramatic is likely to be the fact that the year of the invasion of Iraq, 2003, marked the fiftieth anniversary of three other developments, all of which had a closer relationship to the invasion than might at first be apparent. The first two of those events involved beginnings—the passing of two pieces of legislation in the early days of the Eisenhower administration that established the legal framework for offshore oil drilling. The third involved an ending—the end of nearly a century when one dominant oil-producing nation single-handedly provided more than half of the petroleum in the world.

That nation was the United States of America.

Half a world away from Iraq, just a few months before the start of the invasion, a headline in the *New York Times* had referred to a different kind of battle, and a different kind of risk from petroleum. This second and less dramatic "Gulf war" took place in a different Gulf—the Gulf of Mexico—and it had more to do with tankers than with tanks. In this second set of Gulf battles, a much smaller army was working comparably hard, pitting its wits and investment capital against the elements and the odds. The front lines for this army were located hundreds of miles from the United States, off the southern edge of the continent, searching for weapons of mass *consumption*, in the form of oil. Despite the fact that this search was taking place far from land, the oil deposits were technically "domestic," because the United States had claimed the sea bottoms as part of its "Exclusive Economic Zone." As the *Times* headline noted, however, while this oil was domestic, it was also "Deep and Risky." It was more than a half-mile deep, to be more precise—and that was just the depth of the water. The drill bits would need to drill through additional miles of muck and rock before—if all went well—the effort would finally hit petroleum paydirt. The BP blowout, to note the obvious, would later show what could happen if things did not go so well.[1]

A continent away from the Gulf of Mexico, and another world away from the battles going on in both Gulfs, still another battle was taking place beyond the northern edge of the most remote outpost of the United States—along the Arctic Ocean, north of Alaska. On March 19, 2003, when the second President Bush announced that American and coalition forces were "in the early stages of military operations" in Iraq, few if any television cameras were focused on this third battle. The action taking place in this forbidding region would have been difficult for television audiences to see, in any event—given that it was taking place so close to the north pole, much of the action was going on, literally, in the dark. When Secretary of State Colin Powell made his case for the Iraq war at the United

Nations, on February 5, 2003, he did so only about two weeks after the first sunrise to have squeezed its way above the horizon in Prudhoe Bay, Alaska—the starting point for the Trans-Alaska Pipeline—in the previous two months. Even more than was the case in the sands of Middle Eastern Gulf or the swells of the Gulf of Mexico, the troops that were at work above the Arctic circle were engaged in a battle with the elements, braving even "daytime" temperatures that were about as far below zero as most Americans would have been able to imagine. Other risks in this region included the fact that any television crews actually present almost certainly would have been outnumbered by the polar bears. Save for the Inupiat who have considered this region their home for thousands of years, almost no Americans would have had much desire to be anywhere close to this particular battle, especially during the winter, unless they were forced to be here.

But perhaps that is precisely the point.

In a very real sense we *are* "forced to be" in such forbidding locales. To understand the reasons—and to think realistically about what directions we might want to be considering for the future—it is helpful to consider how we came to move off the edge of the continent in both directions. It is also helpful to recognize the connections to the decisions that led us to move massive military force, once again, into a region of the world where U.S. tanks—whether we are speaking of military tanks or oil tanks—are not likely to be met with cheering throngs of happy civilians.

Two reasons are particularly important, and both of them will be spelled out in greater detail in the pages that follow. One is that the United States simply uses too much oil, too wastefully. The other is that, by the later days of the twentieth century, we had already used up the vast majority of the rich petroleum deposits we once had. Those are the key factors that have led so many brave soldiers of the oil industry to be looking for oil in the realm of the polar bears, or in the deepest oceans ever to be probed by oil drills—to say nothing of the factors leading so many of America's more literal soldiers to find their lives at risk in the sands of Kuwait, or Iraq. They are in such forbidding spots because we are so desperate to find more oil, and we have already used up most of the supplies that are easier to find.

Despite our habit of referring to oil "production," the reality is that the twentieth century was an unprecedented exercise in oil "destruction." The oil was actually *produced* during the time of the dinosaurs. What we have been doing over the last century or more has been to find the fossil deposits left behind during the era of the dinosaurs and to burn them up as fast as we could. Over the course of the past century, we showed an impressive increase in our ability to find those ancient remains, but we didn't manage to create as much as a single barrel of truly "new" petroleum supplies to make up for the supplies we were burning up.

Yet there is also a reason that is significantly less obvious. Our expectations for the future continue to be shaped by the exuberance of the past. That is part of the explanation behind politicians' continued calls for U.S. "energy independence"—generally put forth with straight faces and apparent conviction—when in fact the evidence clearly shows that no such future will ever again be possible, at least not with petroleum. Another part of the explanation for the politicians' continued calls, however, is that the rest of us allow them to get away with it. Perhaps part of the explanation for that, in turn, is that all of us may have some resemblance to the wildcatters who will be discussed in the later pages of this book. We seem to have become so caught up in the excitement of oil strikes that we've started to share the wildcatters' conviction—surely, there must be even more spectacular oil finds out there, perhaps just beyond the next horizon. The problem, unfortunately, is that we are not actually looking toward the next horizon. Instead, we are driving with our eyes fixed firmly on our rear-view mirrors.

All of which means that we are entering a new era in more ways than one. In an earlier century, the United States actually did enjoy something like "energy independence"—or even "energy supremacy"—but as we move into the twenty-first century, any hopes for a "return" to such presumably happy days have less to do with realism than with self-delusion.

The two of us have been studying energy issues in general, and offshore oil issues in particular, for more than thirty years. Near the start of that time, in 1974, President Richard Nixon said, "At the end of this decade, in the year 1980, the United States will not be dependent on any other country for the energy we need." Back then, the United States got 36.1 percent of its oil from foreign sources, and Nixon proposed to end that dependency by obtaining more oil from U.S. sources, particularly offshore oil. The next year, with an emphasis on nearly the same policies, President Gerald Ford said, "We must reduce oil imports by one million barrels per day by the end of this year and by two million barrels per day by the end of 1977." By 1979, President Carter was beginning to place at least some emphasis on different policies, but he made a similar promise: "Beginning this moment, this nation will never use more foreign oil than we did in 1977—never." By that time, the United States was obtaining 40.5% of oil from foreign sources.

President Reagan overturned many of Carter's policy initiatives, particularly those that had to do with solar power and energy efficiency, but he agreed that "the best answer is to try to make us independent of outside sources to the greatest extent possible for our energy." For President Reagan, apparently, the "greatest extent possible" meant importing 43.6 percent of our oil from foreign sources. By 1992, 47.2 percent of our oil was coming from foreign sources, but undaunted, President George H.W. Bush announced that the first principle for his national energy strategy was "reducing our dependence on foreign oil." By 1995, President Bill Clinton said, "The nation's growing reliance on imports of oil ... threatens the nation's security"—and his proposed solution was that we should "continue efforts to ... enhance domestic energy production." At that point, the U.S. was obtaining almost half of its oil (49.8%) from foreign sources. By 2006, the fraction of oil coming from foreign sources had reached nearly two-thirds—65.5 percent—but President George W. Bush confidently predicted, "Breakthroughs ... will help us reach another great goal: to replace more than 75 percent of our oil imports from the Middle East by 2025." By 2009, with 66.2 percent of the nation's oil coming from foreign sources, President Barack Obama announced, "It will be the policy of my administration to reverse our dependence on foreign oil while building a new energy economy that will create millions of jobs."[2]

All of these well-known politicians, and many others, spoke eloquently of the need to promote increased U.S. oil production, to restore the nation's energy "independence." Unfortunately, anyone who actually believes that it would be possible for the United States to achieve anything even remotely resembling "energy independence" would have to be living in a world of nostalgia and denial. U.S. energy independence has not been physically possible since the days when Elvis was still singing—in his truck, not in his recording studio—and if we are thinking in terms of oil, it will never be possible again.

By the time of the *Deepwater Horizon* disaster, decades of policies that were supposedly promoting "energy independence" had left us in deep water in more ways than one. Our efforts to "enhance domestic energy production"—also known as draining America first—have been so "successful" that we are now sending roughly a billion dollars a day to other countries, a number of which don't like us very much, and some of which use their money to attack us. In an ironic twist, the United States may well send more money to fanatical terrorists than does any other country. That may not be the intent, but each one of us may be helping to send a bit more cash to the terrorists each time we fill our gas tanks.

One famous definition of insanity, thanks to Albert Einstein, is to keep doing the same thing, hoping that the results will turn out differently next time. The reality, regrettably, is that even before the explosion of the *Deepwater Horizon*, energy policy experts in the United States have spent decades in continuing to do the same thing, and we, the people, have done next to nothing to reverse the pattern. Instead, we have all been part of a process in which we keep digging ourselves into an ever-deeper hole.

What has happened to date is worth considering in some detail, because that can tell us how we came to be in this hole in the first place. The question now, however, is what we will choose to do in the future. The explosion of the *Deepwater Horizon* provides, in the most vivid form that any of us would ever want not to see, not just a tragedy, but also a challenge, and an opportunity—a challenge to take a closer, more clear-eyed look at our policies, and an opportunity to realize that this is a hole that cannot be escaped simply by digging deeper to look for more oil. Instead, our only hope for a better energy future is to respond to the oil-darkened waters with clearer thinking—to move now to confront the reality of using ever-increasing quantities of scarce and precious petroleum, and to begin the move to a future that will be controlled by our decisions, not by our dependence on the fast-disappearing remnants of the time when dinosaurs last roamed the earth, a good hundred million years ago.

It's about time.

2 The Macondo Mess

Officially speaking, BP's drilling was taking place in a location that the U.S. Department of the Interior calls "MC 252"—government-speak for "Mississippi Canyon block 252." In practice, though, most drilling operations are remembered through their code names, which simplify the protection of confidentiality during early stages of exploration, then provide easier-to-remember names later on. Some of the names come from people (Holly, Heather), others from drinks or fish (Cognac, Marlin), and still others from cartoon characters (Rocky, Bullwinkle). The name of the ill-fated BP project, though, first became known through one of the most important literary works of the twentieth century. It is the fictional town at the center of Gabriel García Márquez's novel *One Hundred Years of Solitude*—Macondo.

One wonders if BP really thought about the implications of that name. The Macondo of the book started as "a village of twenty adobe houses, built on the bank of a river of clear water that ran along a bed of polished stones, which were white and enormous, like prehistoric eggs." So far, so good: BP's Macondo well started in an area of clear water, and although it was looking for prehistoric leftovers in the form of oil, not eggs, that could be seen as a petroleum geologist's version of literary license. If there is any one sentence in Marquez's book that summarizes the fate of Macondo, though, it would be this one, which comes many chapters later: "It was as if God had decided to put to the test every capacity for surprise and was keeping the inhabitants of Macondo in a permanent alternation between excitement and disappointment, doubt and revelation, to such an extreme that no one knew for certain where the limits of reality lay."

In that phrase, García Marquez might well have been discussing the people of the Gulf coast. For at least the first three months after the blowout, the humans along the Gulf of Mexico were in the grip of "excitement and disappointment," unable to know "where the limits of reality lay." For decades into the future, as well, there will continue to be debates over where the limits of reality may be. Still, some of the basic outlines were starting to become clear even as the oil was still erupting into the Gulf of Mexico.

On the day when the *Deepwater Horizon* commemorated the fortieth anniversary of Earth Day by going down in flames, Coast Guard personnel on the scene reported no evidence of spilling oil. In a way, they shouldn't have expected to see much of a problem. Oil companies had been drilling in the Gulf for more than half a century by then, and the last time there had been a significant spill from drilling in U.S. waters—as opposed to the transporting of oil, as in the case of the *Exxon Valdez*—had been the 1969 blowout off the shores of Santa Barbara, which had been part of the inspiration that led Senator Gaylord Nelson to call for that first Earth Day.

Even when truly major hurricanes had ripped through the thousands of offshore oil facilities in the Gulf of Mexico, moreover—Ferdinand in 1979, Andrew in 1992, Ivan the Terrible in 2004, Katrina and Rita in 2005—the underwater shut-off valves known as blow-out preventers had usually done their jobs. After Andrew, when the two of us went out into the Gulf on a research vessel with a group of other scientists, we saw one jack-up rig that had been blown more than a hundred miles away from its original location, but there were no reports of oil leaks even from a force that powerful.[1]

In the case of the *Deepwater Horizon*, unfortunately, oil did soon begin to appear. Initially, there was some hope that this oil might just have come from the mile or so of pipe that connected the rig to the ocean floor. Over the next few days, unfortunately, it became clear that officials were seeing far more than just one pipe's worth of crude oil. By April 30, the headlines in the *Mobile Press-Register* called attention to a leaked report, "Government fears Deepwater Horizon well could become Unchecked Gusher." Soon thereafter, the fears were confirmed by growing quantities of oil itself.[2]

There is no good time or place for an oil spill, but it would have been hard to do much worse than the springtime spill at Macondo. The ten-year-long Census of Marine Life, concluded in 2010, named the Gulf of Mexico as the fifth-most-diverse marine setting in the world for known species. Not only did the deep-water Mississippi Canyon location have a significant ecological sensitivity of its own, but there were at least three other ecological worries nearby. First, the spill took place only a few miles from what oceanographers call "the Loop Current"—a massive conveyor belt of warm water in the Gulf that squeezes northward through the gap between Cuba and Mexico's Yucatan peninsula, coming very close to the spill location, before heading to the east and then south, looping back down below the tip of Florida and then heading up the east coast under the better-known name of the Gulf Stream. Second, the entire Gulf region sits beneath one of the world's major migratory flyways—a virtual superhighway for millions of birds heading north and south—as well as providing irreplaceable habitat for other birds that spend most of their time in the region, including herons, egrets, and over a third of the nation's brown pelicans. Third, and perhaps most seriously, the shorelines nearby included not just beautiful beaches but the highly sensitive habitat of the Louisiana wetlands—3.5 million acres of coastal wetlands, or about 40 percent of all of the coastal wetlands in the continental United States. Adding to the injury, BP was spewing out its oil during the very times of the year when many species, from brown pelicans to bluefin tuna, were breeding, spawning, hatching, and fledging, exposing the young of the species to maximum threat at the very times of their lives when they were the most vulnerable.[3]

The top officials of BP initially did all they could to minimize the damage, or at least the public relations damage. Their initial estimate of the amount of oil being spilled was an even 1,000 barrels per day—a quantity that led BP's CEO, Tony Hayward, to note that "the Gulf of Mexico is a very big ocean, and the volume of oil we are putting into it is tiny in relation to the total volume of water." In percentage terms, or to him, that might have seemed accurate, but the spill size estimate was not. Within days, an independent scientist used satellite images to estimate the spill volume at 5,000 barrels a day. After initially disputing the independent estimate, BP eventually agreed with it—but others did not. By mid-May, National Public Radio would report the results of other independent scientists' estimates, most of which were around 50,000 barrels per day—about on par with the estimates we had been hearing informally from industry insiders, more or less from the start, based on production volumes from other oil wells in the same area. By late June, Marcia McNutt, the Director of the U.S. Geological Survey and the leader of the federal government's "Flow Rate Technical Group," estimated that the blowout was spewing between 35,000 and 60,000 barrels per day.

In the refined estimate that was released by the Deepwater Horizon Unified Command as the leak was finally being plugged, government scientists concluded that their best estimate of the total spill near the upper end of that range—starting at 62,000 barrels per day, but still spewing at the rate of 53,000 barrels per day just before the closure, for a total spill volume of 4.9 million barrels.[4]

BP's original estimate, to put it kindly, was a bit on the optimistic side. It was only about 2 percent of the actual volume, which was actually around sixty times higher. At the 60,000-barrel-per-day rate, the terrible torrent would have been shooting about 2.5 million gallons of petroleum into the Gulf every day, or more oil than another *Exxon Valdez* spill every four and a half days. Given that a barrel of oil contains 42 gallons, that came to more than 200 million gallons of spilled crude oil. The BP blowout thus became the largest peacetime offshore oil spill in history, passing the massive 1979 spill of Ixtoc I, off the shores of Mexico, as well as the onshore, "Silent Spill" at Guadalupe, California, which went on for more than four decades before finally being stopped during the 1990s.[5]

It was not, however, the biggest spill ever. Near the end of the war in Kuwait, Saddam Hussein had managed to spill more oil than escaped from Macondo, but that wasn't easy. He put his national army to work, deliberately opening the oil spigots of Kuwait as his forces retreated in 1991, setting fire to some 700 oil wells, and surrounding the burning wells with land mines to keep the firefighting crews away. Hussein, however, was trying to spill the oil; BP, as we were told repeatedly, was trying to shut it off.

The clean-up and containment efforts, unfortunately, proved to be significantly less impressive than the quantities of oil gushing out. BP initially presented an optimistic picture of a company scrambling to clean up the mess, mobilizing a "flotilla of vessels and resources that includes: significant mechanical recovery capacity." By the start of July, however, with oil still spewing into the Gulf at a stunning rate, the *Washington Post* noted that BP's "significant mechanical recovery capacity" was actually removing less than 900 barrels per day—about as much as the company's initial, overly optimistic guess at the total spill volume. That meant that, after more than 75 days of supposed containment and clean-up activities, BP had managed to skim a *total* of just 67,500 barrels and to burn off just another 238,000—little more than half of the 491,721 barrels of oil that BP's recovery plan claimed the company would have the capacity to remove *every day*.[6]

All in all, it was not the kind of end that most people would have expected for the *Deepwater Horizon*.

For most of the time while we were working on this book, the battle to control the spill was still raging. By the time we were sending this book to the publisher, that battle had finally turned the corner. The task now is to start looking ahead, beginning by asking a bit more systematically about the lessons we need to learn.

That point applies with particular force to the lesson that will be the major focus of this book. Both literally and figuratively, and both in the Gulf of Mexico and elsewhere, we have been getting into increasingly dangerous waters without being sufficiently vigilant about the implications of our actions. It is a theme that will show up in three main variations—in terms of the risks being posed by present-day oil-drilling technologies, the apparently economic choices that our political system has been putting into place over the decades, and the broader energy choices that all of us have been making, sometimes without thinking very hard about it, for our overall ways of life.

The issue of technological risks, which is the one that was the early focus of policymakers and the press, is also in some ways the most straightforward, and so it will come first. The *Deepwater Horizon* shares many of the characteristics of other technological disasters, including the fact that, at least in retrospect, observers can easily spot the warning signs that were there to be seen—some of them subtle, but others truly flagrant, and almost all of them in plain sight—well in advance of the blowout. At the same time, though, it is important to recognize that this was no third-rate technology.

The management and the regulation of the technology, on the other hand, may indeed have been third-rate, or even worse. BP and its contractors had taken a series of cost-cutting moves, including the fact that the *Deepwater Horizon* had been officially registered as a ship under the flag of the Marshall Islands, limiting the potential ability of the U.S. government to regulate its operations. In the aftermath of the blowout, many Americans could well have wondered if it would be possible for the standards of other countries to be even weaker than those of the United States of America, but "different types of rigs are classified differently, and the Marshall Islands assigned the Deepwater Horizon to a category that permitted lower staffing levels" than U.S. agencies would have required.[7]

For the most part, though, the hardware itself was close to being the best we had, at least in the United States. This was the very rig that held the all-time record for deepwater drilling at the time of the disaster, and few other drilling rigs in the world were more sophisticated than the one that now lies crumpled at the bottom of the Gulf. The technology that was in use, including the blowout preventer that failed to do its job, was officially considered to be "fail-safe" just a week before. Unlike the supposedly "unsinkable" Titanic, the technology being used by the *Deepwater Horizon* had long been used in many other locations, and although it clearly wasn't fail-safe, it had generally seemed to work before. The vessel had even been inspected just a few days before disaster struck, and at least according to the records of the agency in charge, the U.S. Minerals Management Service, it passed with no problems.[8]

That may be true bad news.

Still, that may also be the right place to start in making sense of this disaster—understanding how it happened, and what kinds of changes we need to start making now, to make sure that such a thing is not allowed to happen again. The fact that the blow-out "preventer" was officially seen as fail-safe, for example, can have important implications for future decisions if we start to ask the right questions about it today. Even at this early stage, though, it is clear that this disaster was not simply due to an imperfect piece of hardware. Although the blow-out preventer failed to work, the humans who were responsible for drilling clearly left even more to be desired—a point that extends not just to BP and its contractors and partners, but also to the regulatory system that was supposedly keeping a close eye on things.

As more bad news began to leak out, along with the oil, it became increasingly clear that the rig operators—and the regulators who were supposed to be watching them—had in fact often been cutting corners on safety. The flood of oil will be followed by a flood of income to lawyers, as they argue over the details of this point, but based on the evidence available to date, it appears that the cutting of corners did contribute directly to the disaster. Still, it is important to recognize at the same time that this disaster occurred in a system that, at least a week before the spill, was officially considered to represent the state of the art. The problem is, as the Macondo blowout made clear, the state of the art wasn't good enough.

In the thousands of lawsuits that have already begun to be filed, as well as in those that are still to come, much of the focus will be on the economic and other losses being experienced by the ecosystems of the Gulf coast, and importantly, on the people who have long depended on those ecosystems for their ways of life. The news on that front may not be exactly heartwarming. Ironically, although the last major oil spill in U.S. waters had taken place more than twenty years before the BP gusher, the experience with that spill—the grounding of the *Exxon Valdez*—was that checks were only starting to be made out to many of the "successful" litigants in the lawsuits a few weeks before the *Deepwater Horizon* blew up. Before the checks started flowing, the unfortunate victims of that earlier spill were subjected to levels of stress—including stresses related to the litigation process itself—that few of them could have imagined in advance.

By the time the checks were finally being written, one out of every five of the "successful" victims in the lawsuit had died. Of those who were still alive, one out of four faced liens on the checks they would finally receive. One Alaskan we know said that the check, while welcome, wouldn't be enough to cover the cost of a bankruptcy filing. If the *Exxon Valdez* is any guide, many of the fishermen of the region who are currently hearing promises about being "made whole" by the "responsible" parties will find those promises—and their bank accounts—to be just as empty as their fishing nets now are. For others in the region who are also suffering losses due to the disaster—owners and employees of beauty shops, restaurants, and a thousand other small businesses that are no longer earning a living from fishermen or tourists—the odds are probably even worse.[9]

The lawsuits, however, are by no means the only arena of economics where it is important to take a closer look at the implications of our choices.

As a matter of law, the oil that came pouring out of BP's ill-fated Macondo well belongs to us, the people—the 300+ million citizens of the United States of America. BP and other oil companies have been allowed to remove that oil, and to enjoy the profits from doing so, under a system in which the federal government "leases" the drilling rights, collecting royalties and other fees in return. In some ways, the amounts of money involved are huge—over the past half-century, the federal program that leases the oil has been second only to the Internal Revenue Service as a source of income for the federal government. In other ways, however, the amounts are amazingly small. According to the U.S. Government Accountability Office, the United States gets a lower share of the income from our nation's offshore oil than almost any other jurisdiction in the world. The total "take"—including leases, royalties, bonus payments, and even the income taxes that corporations pay on the profits they make from the people's oil—is lower in the United States than in almost any other nation, anywhere. Several U.S. states, including Colorado, Wyoming, Texas, and even Louisiana, get more tax revenue than the U.S. government. For deepwater leases such as the one where the *Deepwater Horizon* was working, U.S. taxpayers get about 40 percent of the "take." Norway charges almost twice that much—about 75 percent. Vietnam and Tunisia get an even higher share—80 percent or more. So do Angola, Kazakstan, and Brunei.[10]

The United States, in short, has scarcely been the shrewdest business operator in the world. The oil corporations have been enjoying awe-inspiring profits, and thanks to the U.S. government, you and I and our fellow taxpayers have all helped to make those profits possible—whether we buy gasoline or not.

In recent years, moreover—as world oil prices and oil company profits have soared—Congress has chosen not to bring U.S. revenues any closer to those in the rest of the world.

Instead, Congress has taken actions that our representatives have assured us were in response to sound energy policy concerns, and not to the fact that the oil industry's contributions to its "friends" in Congress have been as generous as the taxpayers' share of the take has been puny.

In 1995, for example, Congress chose to *cut* the already-low fees that the U.S. was charging the oil companies, passing the Outer Continental Shelf Deep Water Royalty "Relief" Act. Ten years later, the 2005 Energy Policy Act extended other favors. The committee chair at the time, Rep. Joe L. Barton (R-Texas)—a friend of the industry who would make news again in the aftermath of the BP blowout by apologizing to BP officials for the harsh treatment they were receiving from the Obama administration—made last-minute additions of billions of dollars in tax and royalty "relief," to encourage faster drilling, along with a $50-million annual earmark to support technical research for the industry. All in all, the Government Accountability Office has estimated that the deep-water royalty waiver program, alone, could cost the nation more than $50 billion over the life of the leases.[11]

Still, in some ways, the most important version of this lesson may be the biggest one. It is not just along the Gulf coast that Americans have become dependent on a resource that is rapidly disappearing. It is not just in the case of the *Deepwater Horizon* that our society has come up against the limits of our technology, or perhaps beyond. In broader ways, as well, we are coming face to face with a larger, technological Peter Principle.[12]

You may remember the original Peter Principle—the idea that, in an organization, someone who seems competent is likely to get promoted, and to keep getting promoted, until reaching a job where he or she is no longer competent. Since that's where the promotions stop, the argument goes, each individual rises to his or her own level of incompetence in an organization, then stays there.[13]

The problem that should have become too obvious to ignore, with each new barrel of oil that came shooting out of the ocean floor, is that the technological version of the Peter Principle may not be nearly as humorous as the original statement. Instead, we may have "promoted" the societal significance of our technology up to, and perhaps beyond, the point where it can actually do what we expect it to do.

As the following pages will spell out, the result is easy to describe, but ominous to consider: Not just in the Gulf of Mexico, but across our technology more broadly, our capacity to *do* damage, both to ourselves and to our environment, may well have risen faster than our capacity to *un*do the damage. Meanwhile, rather than confronting the realities of growing ever more dependent on a form of fuel that is made out of ancient fossils and is disappearing faster every day, we have allowed our politicians to offer platitudes about "energy independence" that, by now, no one with an IQ score in positive numbers should be able to repeat with a straight face.

To note a figure from the *Wall Street Journal*—not normally a source of excessive alarm about the limits of capitalism— the world now uses up about a thousand barrels of its finite oil reserves every second. The United States has long been burning through that oil faster than any other country, even though we have only 5 percent of the world's population. The same nation was for a very long time also the *source* of most of the world's oil, but today, the United States has only about 2 percent of the world's remaining proven reserves. With each barrel that spews into the Gulf, *and* with each barrel that disappears out the tailpipes of our cars, less of that finite oil is left. That fact, it seems, deserves a bit more attention than it has received to date.[14]

3 Stored Sunlight and Its Risks

As we think about that rapidly disappearing oil, it might be helpful to have at least an English-language understanding of how it got to be there in the first place. The basic starting point is that, like other forms of fuel, oil and gas can be thought of as a special form of stored sunlight, or more specifically, photosynthesis—the process through which plants use sunlight to produce organic material, such as leaves, straw, wood, or cellulose. Oil, gas, and all other fossil fuels are left-over remnants from plants, from animals that ate the plants, and from the animals that ate the animals that ate the plants, all of which have long been stored in special forms of sedimentary rock.

As the name implies, sedimentary rock was first formed through deposits of sediments—the stuff that settles at the bottoms of water bodies, or in some cases gets carried around and then dropped by wind. Sedimentary rock is different from what geologists call *igneous* rock—the name of which is derived from the Latin word *ignis*, meaning fire—which was instead first formed through the cooling of molten rock, whether we are talking about an ancient molten blob that formed the earth, billions of years ago, or a spurt of volcanic lava next week. Some types of sediment, such as mud, sand, silt, or pebbles, are formed as igneous rocks are broken down by weathering.

Other sedimentary deposits were formed in spots where salt was precipitated out of sea water, settling to the bottom. Sediments may also be produced from organic leftovers, such as the sea shells that settled out on lake and sea bottoms many centuries ago. As early sediments come to be buried under newer ones, the pressure that comes from being buried more deeply can cement the individuals grains of sand or silt into layers of rock.

Matters become interesting for petroleum geologists in spots where organic matter—the remnants of that stored sunlight, ranging from microscopic organisms to decomposed dinosaurs—is mixed in or under such sedimentary deposits. Organic matter decays as it mixes with oxygen, so if the deposits are exposed to the atmosphere, much of the organic matter will turn into gases, escaping through the pores in the rock. In cases where the organic matter is cut off from supplies of oxygen, however—as for example where deposits are quickly covered by new sediment, or where the deposits pile up in deep, oxygen-free waters—some of the organic matter can remain embedded in the sedimentary layers that form.

Over time, as new deposits are added and the original sediment is compressed into rock, the embedded organic material gets squeezed into pores or *voids*—tiny openings between and among all of those individual grains of sand and silt, being fused together within layers of rock. Muds, sands, and silts form sandstones and shales, while organically produced sediments such as seashells form limestones. As the rocks are buried deeper, the heat increases, and at about 150°F, the organic material starts to be "cooked" into a homogenous liquid. This is the key point where the substance we call oil begins to form.

Depending on a variety of factors, such as the type of organic material, the type of sediment, the temperature, and length of time this pressure cooking goes on, the resultant fluid can take any number of colors, ranging from almost clear to deep black in color, with greenish, reddish, and brown shades being common variants in between. The liquid can also range from being thin to very thick in viscosity. One variation, however, is particularly notable. As the temperature increases, gas begins to form, and as the temperature rises past 300°F, oil is irreversibly turned into gas and carbon (graphite). One implication is that, in general, the deeper the deposit, the higher the gas-to-oil ratio will be.[1]

For reasons that are not hard to fathom, the deposits containing organic materials that have been converted to oil or gas are often called *source rocks*. For commercial quantities of oil or gas to be present, however, it is not enough simply to have source rock. Instead, two other conditions must also be met. First, there must be a reservoir rock, or a permeable layer of rock that can store the oil or gas. Second, there must be a *trap* in the reservoir that concentrates the petroleum and keeps it from escaping.

Again speaking in general terms, oil and gas can be expelled from the source rock by either of two processes. First, as the source rock gets buried deeper and deeper, the rock compresses, leaving less space in the pores between the sediments, and squeezing out the oil, just as water can get squeezed out of a sponge. Second, the creation of oil out of unconsolidated organic matter can cause an increase in the total volume of that organic matter, which can fracture the source rock, allowing the oil to escape. In either case, since oil is lighter than water, it will migrate upward until or unless it is trapped by a cap on top.

As a general rule, sedimentary rock starts out as a flat layer—think of the mud or sand that winds up at the bottom of a small lake. If oil is uniformly distributed in such a layer of reservoir rock, it is very hard to extract by drilling. An example is provided by the *oil shale* of northwestern Colorado—a rock that contains significant quantities of petroleum, and that will actually burn if it is exposed to a flame for long enough, but that has resisted any efforts to date to remove the oil at a reasonable cost. Instead, the oil or gas must be concentrated—something that occurs as the oil and gas migrates into traps in the rock. Oil, gas, and water are all present in reservoir rock, but the oil will not migrate unless there is a change in that rock, such as bending or fracturing.

Imagine a flat layer of reservoir rock, with 5 percent of the pores filled with oil and gas, and the remainder filled with water. Over geological time scales—meaning millions of years—powerful forces can squeeze and twist the rock formation, putting tremendous pressure on that reservoir rock, and on the layers of sedimentary rock around it, forming huge wrinkles. Since oil and gas are lighter than water, they will squeeze upward, and if there is a trap, they will tend to collect or concentrate near the tops of the wrinkles. The upper part of a given wrinkle can become an oil and gas reservoir, "floating" on top of the water below, and offering a promising spot to look for concentrated petroleum. In general, any gas in the reservoir will be at the very top, above the heavier oil, which in turn will be above the still-heavier water. Think of what would happen if you fill a glass half-way with cooking oil, then turn it upside-down as you plunge it into a kitchen sink that is already half-filled by water. The air would be trapped on top, under the glass, the cooking oil would be trapped in a layer below the air, and water would stay below—offering a reasonably good analogy for the kinds of traps that petroleum geologists hope to find.

There are many ways for sedimentary rock to form, because earth has always been geologically and climatically active. The southern portions of the American Great Plains, for example, were once an inland sea, where layers of dead shellfish were deposited, resulting eventually in what we now know as the oil reservoirs in Oklahoma and Texas. In earlier geologic times, similarly, the shoreline of the Gulf of Mexico was much further out than it is today, and ancient ancestors of the Mississippi River deposited countless layers of sediments as they built deltas—in places that include Mississippi Canyon block 252, where BP's Macondo well blew out—that are now buried by thousands of feet of other, more recent deposits. Sediments that start at the surface can eventually get buried very deeply, but at the same time, petroleum that is generated in deep geological formations can eventually be found in shallow reservoirs. Mountains rise and buckle, and reservoir rock can tilt so steeply that, given slope and time, the oil and gas can migrate great distances. In the Williston basin of Montana and the Dakotas, for example, "the oil has migrated over 200 miles (320 km) from the source area in the deep basin to the traps on the flanks of the basin." Once these ancient reservoirs are concentrated, however, then wherever they wind up, they are at least potentially available for "discovery" by some of the planet's relative newcomers—humans.[2]

Discovering Oil

Early exploration and drilling for oil usually took place where people saw evidence of oil or gas on the surface, most often in the form of oil or gas seeps. The site in Pennsylvania that Edwin Drake and his partners chose for their first effort to drill for oil, for example, was at a place that already bore the name of Oil Creek, and the partners chose that site because of the seeps that gave the creek its name. They managed to succeed by following the logic that, where oil or gas was coming out of the ground, there must be more of it below, and for the next half century, most drillers continued to follow the same approach.

Not until the early years of the twentieth century did it become increasingly accepted that the places to find oil and gas would be in upper layers of certain sedimentary rocks. It was around that time that geologists began to map outcroppings of these layers on the surface, then working backward to identify likely traps below.[3]

The ability to start "seeing" such formations beneath the surface—or more accurately, "hearing" them—would come along another two decades later, in the 1920s, with seismic technology. Seismic exploration works with sound—much like a depth finder on a fishing boat. Bursts of low-frequency sound are sent downward, and the depth finder calculates the time it takes for the sound waves to bounce off the bottom and return to the surface. On the boat, the time calculation is then translated into an estimate of the water depth; in the search for oil, the intention is to map the shape as well as the depth of the underground wrinkles. For the early use of seismic technology in the oil industry, the sound was generated by an explosion on the ground, or in a shallow well; technologists could then use a series of recorders, or *geophones*, arranged in a grid, to find out how long it took for the sound to return to different spots on the surface. With this technology, it was possible for the first time to start mapping the subsurface rock formations, noting such things as domes or the slope of a given layer of rock. Even though these early seismic technologies were relatively primitive by today's standards, they could be coupled with surface mapping of rock outcroppings to give improved indications of where to find promising traps for oil.

The technology has evolved considerably over the years, but an especially notable advance involved the use of increasing computational power, particularly in the 1980s, which led to the emergence of three-dimensional seismic exploration. Using many more recorders, complex algorithms, and the ever-increasing processing power of more advanced computers, it became possible to generate far more detailed maps of subsurface rocks. In recent decades, the literal explosions that were used in earlier seismic explorations have become increasingly rare, having been replaced by the use of vibrator trucks on land, and air guns on water, to generate the initial sound signals. The vibrator truck is a heavy vehicle with a large pad in the center, which can be lowered with hydraulic pistons to take up most of the weight of the truck. Hydraulic motors then use the weight of the truck to shake the ground, providing the sound waves that are needed for analysis. Air guns are cylinders that are towed behind a survey boat, 20 to 30 feet below the surface. At precisely timed intervals, ports in the cylinder are opened, and high-pressure air is released to produce the ground-penetrating noise signals. A seismic array several miles long, with spaced geophones, is towed behind the boat to record the returning echoes.

Finding a promising spot to look for oil, though, is only part of the process. There is also that small matter of drilling down to it and getting it out—safely.

Getting the Oil Out

The basic techniques for drilling an oil or gas well are fairly straightforward, but the details and the accomplishments can get increasingly tricky as the environment becomes more challenging. The very first well in Pennsylvania used the techniques and technology of a local salt well driller, involving a wooden frame and a drill bit—basically a steel rod with a reasonably sharp point—that was pounded into the ground by human labor. It was not the kind of enterprise that anyone would want to use for a well that could be thousands of feet deep. Today's drilling bits rotate, instead of being pounded, but they don't look or work like the drill bits you can buy in a hardware store. Instead, they have several sets of teeth, and they rely on a complicated set of tubes and casings.

The basic reason is easy enough to understand—after the drill bit cuts through rock, the rock doesn't simply evaporate. If you have ever drilled a hole through a relatively thick board, you may remember that you needed to stop a few times, pull out the drill bit, and clear out the "sawdust" (actually chips and shavings) before you could start drilling again. If you were to be drilling through solid rock, on the other hand, a mile or more below the surface, you wouldn't want to be pulling the drill bit all the way out of the hole every time you drilled a few inches deeper. Instead, drillers have developed an ingenious system to get the job done in a more efficient way.

In essence, an oil-drilling rig is a machine that is capable of lifting, lowering, and turning heavy segments of pipe, as well as pumping drilling mud and cement. The starting point is what the industry calls a *drill string*—a set of interlocking, heavy-duty pipes with a drill bit on the bottom—and a mechanism on the top for turning the string and the bit. The early bits were little more than steel shanks with flattened and sharpened heads, but as wells got deeper, more sophisticated bits and techniques were developed. Today's drill bits have embedded carbide cutting surfaces, and unlike the bits from a hardware store that have grooves on the outside, the oil-drilling bits have a hole that runs through the center. As the bit cuts deeper into the earth, new sections of drill string are added at the top.

Early in the process, a larger pipe known as a *casing* is lowered into the well, to serve as a liner along the side of the well. At that point, the well is temporarily plugged, below the casing, and cement is pumped down the well. The plug below the casing forces the cement back up between the casing and the geologic formations through which the well is being drilled, cementing the casing in place, and sealing off the well from the surrounding soil and rock. Two bits of terminology will be helpful to remember here. First, the gap between the casings and the rock is called the annular space, and second, the important piece of equipment that is installed on the top of the first casing is called a *blowout preventer* (BOP)—a large valve or series of valves that, if all goes well, should be able to seal off the well in the case of an emergency. In offshore drilling, incidentally, the top of the BOP must be connected to the drilling rig with a *riser*, which is a pipe that allows the circulation of *drilling mud* from the rig through the well—more on this in a moment. In the case of BP's Macondo well, this riser was almost a mile long.

After the first casing is put in place, a slightly smaller bit is installed; the new bit and drill string are lowered through the BOP and casing, and the drilling continues. Eventually, another casing—again a slightly smaller one—is lowered into the well. The top of this casing is connected to the bottom of the previous one, and then it, too, is cemented in place. Through a series of smaller casings, the well proceeds downward, starting to look a bit like a collapsible or multisegment telescope, with the eyepiece or small end at the bottom. Unlike a telescope, though, each segment of an oil well gets longer as the well goes deeper. As the drilling proceeds, each segment is attached to the previous one and then cemented in place, meaning that there should be two ways in which the well is sealed at each segment—one by being attached to the previous segment and the other by the cement that surrounds it.

The reason for the hole in the center of the drill bit, and the drill string, is so drilling mud can be pumped down through the drill string and the hollow tip of the drill bit. Under pressure, the mud squeezes out around the drill bit at the bottom, flowing back up to the rig through the space between the drill string and the casing. Two of the reasons for using the mud are similar to what you might do while drilling a smaller hole—pulling out the bit to get those shavings out of the way, and possibly blowing on the bit to cool it off. First, the mud collects what drillers call *cuttings*, or the small pieces of rock that have been cut by the drill, carrying them back up to the surface and out of the well. Second, the mud lubricates and cools the cutting surfaces as the bit cuts through rock. You might not have needed to stop in drilling a smaller hole if you could have used drilling mud—but more to the point, you might have no way to drill a really deep hole, literally miles deep, without such a substance. What we call drilling "mud," incidentally, is actually a specialized mixture of clay and other substances, often containing barite and a mixture of industrial-strength chemicals, which are so heavy that even rock cuttings will be suspended in the mud. At the rig, these cuttings can be filtered out of the mud, which can then be pumped down through the drill string again.

The third reason for the mud, though, is something that is rarely a concern for someone who is simply drilling a hole through a board. Once the well reaches a reservoir of oil or gas, the mud is critical for controlling the pressure in the well.

As suggested above, most petroleum reservoirs contain some mixture of oil, gas, and water. Although liquids such as oil and water cannot be compressed very much, anyone who has ever blown up a balloon knows that ordinary air, like any other gas, can be compressed quite a bit. Because these reservoirs are often located very deep in the earth, the gas in them can be under far more pressure than any human set of lungs could create—often amounting to thousands of pounds per square inch. Just as pressurized air can cause a balloon to explode if it is punctured by even the tiniest of holes, the pressurized gas in an underground formation can exert an essentially explosive force to shoot out of the well—drillers call it a *kick*. If the kick is not controlled, the drillers can face an extremely dangerous, uncontrolled situation called a *blowout*. Gas pressure was the factor behind the "gushers" that early drillers encountered in some of their shallow oil wells, as recorded in old, black-and-white photos. It was also a key factor behind the volcanic-type eruption of oil that started spewing from the BP well, which Americans were able to see live, on their TVs, during the spring and summer of 2010.

Although other factors are often at work, the usual expectation is that deeper reservoirs will have greater pressure. This is one of the reasons shallow wells usually need to pump the oil to the surface, often using the "rocking horse" pumps that are seen in early oil-producing areas, while deeper wells are more likely to squeeze out a flow of oil without any such pumping, because of internal reservoir pressure. The drilling mud, fortunately, obeys some of the same laws of physics as do the oil and gas. The deeper the well, the more weight there will be at the bottom of the column of drilling mud, counteracting the gas pressure and helping to prevent a well from "blowing out" when the bit penetrates the reservoir. As BP demonstrated, though, while water has at least some of the same characteristics—a mile below the surface, the water can "weigh" about a ton per square inch—mud is thicker than water. While gas can easily bubble through water at almost any depth, the higher viscosity of specialized drilling mud can help to keep the gas confined, at least so long as other conditions are right.

Sometimes, the net result of all that drilling will be nothing more than a *dry hole*—a very deep and expensive "underground air-storage tank," in the words of one long-time oil man, with no petroleum payoff at the bottom. If the drillers are more fortunate, though, they will reach an oil reservoir. At that point, various tests are run to determine the size and potential flow of oil or gas from the reservoir. Onshore, even this initial well may be used to produce oil. With offshore exploration, however, the usual procedure is to drill one or more exploratory wells in a new area, to gather more information first. After testing, such wells are sealed with cement and then "abandoned," except that the drilling company has a strong incentive to keep careful track of the well locations, at least in the case of the wells that hit oil.

Once the initial information provides a better picture of the producing structure and its characteristics, the oil company will design and fabricate an offshore production facility—complete with an impressive set of control mechanisms and valves—after which it will drill *production wells* to extract the petroleum. As you may remember from that upside-down glass in the sink—gas on top, oil next, and water below that—almost all petroleum reservoirs contain water, but it's not the kind of water most of us would want to drink, since it tends to contain various dissolved minerals, usually including salt. Drillers do their best to reach the oil at a level above the water, but there are times when the water will also come up. These *produced waters*, as they are called, need to be treated, discarded, or in some cases, pumped back into the reservoir to maintain pressure.

Since the gas will be above the oil, drillers also try to tap into a reservoir at a level that is below the gas, counting on the pressure from the gas to help push the oil up to the surface. Any natural gas that comes up with the oil can be captured and sold, at least in principle, but many wells are far enough from gas pipelines that the gases are simply burned off, or *flared*. That process, incidentally, has important implications for global climate disruption: Work by the World Bank indicates that the flaring of the natural gas from commercial oil fields "adds about 390 million tons of CO_2 in annual emissions"—more than the sum total of all the projects currently registered under the Clean Development Mechanism of the Kyoto Protocol for reducing greenhouse gas emissions. A more productive approach is one that is used on the North Slope of Alaska and in a number of other oil fields where there is no way to get the gas to a market: The gas is pumped back down into the oil and gas reservoirs to help maintain the pressure.[4]

But the issue of gas pressure also brings us back to the BP blowout and the *Deepwater Horizon.*

Torrent of Troubles

The exploratory well that the *Deepwater Horizon* was drilling on Mississippi Canyon block 252 was in a spot where the water that was almost a mile deep, and the oil reservoir was another two and a half miles below the seabed. BP started drilling an exploratory well in that spot on October 7, 2009, using a drilling rig named *Marianas*. The geology looked promising, but the other early omens did not. On November 9, 2009, barely a month after the drilling started, *Marianas* got battered by Hurricane Ida.

After a delay of about three months, BP and Transocean replaced *Marianas* with the *Deepwater Horizon*—the rig that had set the record for drilling the deepest oil well in history just a month before the *Marianas* had started drilling, having found oil more than 35,000 feet deep, or a mile deeper than Mount Everest is tall. Still, by the time the *Deepwater Horizon* resumed drilling, on February 6, 2010, the well was behind schedule and over budget. According to work on the role of human and organizational factors in risk management, the time and money pressures should have provided more important warning signs than the fact that the first rig got hammered by a hurricane.

In an interview with CNN's Anderson Cooper shortly after the blowout, several BP workers who survived the ordeal reported that, by the time the rig blew up on April 20, drilling was five weeks behind schedule and more than $20 million over the budgeted cost. They reported an argument between BP's "company man" and a representative of Transocean, the owner of the *Deepwater Horizon*, in which the company man specifically said that the operation would skip some of the usual safeguards, and that, "that's the way it's gonna be." Their lawyer, Steve Gordon, claimed that BP's actions displayed "at least" negligence: "I've seen negligence, I've seen gross negligence, and this conduct is criminal.... There's a crime scene sitting 5000 feet below the water." When the five workers were asked which they thought BP considered more important—time and money, or safety—their immediate answer was, "time and money." As one of them put it, the operator "preached safety," and "everybody has the right to call time out for safety. But you do it, you're gonna get run off. You know, you're gonna get fired. They're not gonna fire you for that, but they're gonna figure out a way eventually to get rid of you."[5]

The usual caveats need to apply. The workers in that interview, after all, were suing BP, and their lawyer obviously had an interest in winning as big of a settlement as possible. People in such a situation may well have a temptation to embellish their remarks, and in its own statements, BP has said repeatedly that the company really did put "safety first." Still, based on patterns that have played out all too often in previous technological disasters, and in what the two of us have often seen, through decades of involvement in risk assessment and management, we find the workers' claims to be depressingly easy to believe.

In the wake of the *Exxon Valdez* oil spill in 1989, the field of risk analysis started devoting increased attention to the risks that could be created by humans and organizations, not just by hardware. Part of that new focus included the concept of "the atrophy of vigilance"—an expectation for organizational performance to get sloppier, over time, particularly in the case of rare or "unexpected" problems. The prediction, in brief, was that nearly everyone would show high levels of vigilance in the immediate aftermath of an accident, but that vigilance could be expected to erode, or atrophy, as time went on. It would be an exaggeration to say that this work "predicted" the *Deepwater Horizon* disaster, but it would be right on target to say that the work predicted there would be problems like this one—in part precisely because the organizations involved had managed to pull off earlier forms of corner-cutting without experiencing any obvious negative consequences.[6]

One way to think about the problem is by considering something that almost anyone would consider to be an unwise approach to risk management—Russian Roulette. If there is a revolver with six chambers for bullets, and only one bullet, then if you happen to point that gun at your head and pull the trigger, your "maximum likelihood estimate," to use the risk-analysis jargon, is that you will not actually blow your brains out. Instead, the odds are five out of six, or about 83 percent, that you will be able to take that risk and still survive—at least once.

Most industrial risks are like that, except there is no way to know for sure how many bullets, or how many bullet chambers, are in any given revolver. Professional risk managers warn their colleagues, over and over again, that the prudent thing to do is to remember that the next pull of the trigger, or the next decision to cut another corner, might be the one that will prove to be a fatal mistake. As the risk managers know as well as anyone, though, the reality is that there is no way to know—at least not until it is too late to put the bullet back into the revolver. Making matters still worse, each time an industrial operation manages to cut corners without encountering disaster, there can be more of a temptation to conclude that the risk managers are "just crying wolf," or raising alarms about something that will probably never happen anyway.

An additional part of the problem involves complacency, or even boredom. At least since the time when an iceberg got the best of the "unsinkable" *Titanic*, most ships' crews have shown increased levels of caution and alertness when steaming through iceberg-infested waters—at least for the first time. Adrenaline, however, does not keep on pumping forever. Before the disastrous journey of the *Exxon Valdez*, over 8,000 tankers had gone in and out of the Alaska Pipeline terminal in Valdez, over a period of more than a decade, without a single catastrophe. Although prior shipping had not been totally immune from problems, the general pattern of experiences up through 11:59 p.m. on March 23, 1989, didn't seem to offer any particular reasons for concern.

That absence of concern, unfortunately, seems to have been precisely the mind-set on board the *Exxon Valdez*. Like the *Deepwater Horizon*, or even the *Titanic* itself, Exxon's ship seemed to be state-of-the-art hardware—it was the largest, newest, and best-equipped oil tanker owned by Exxon, which in turn was one of the largest and most technologically sophisticated organizations in history. Five minutes later, however—despite an array of navigational devices with a level of sophistication that early sailors could scarcely have imagined—the high-tech tanker had a spectacularly low-tech encounter. It ran straight into a rock that was clearly noted on all the navigation charts, and that literally had a red flashing warning light on top, sending out a warning that any sailor should have been able to recognize. That rock, officially known as Bligh Reef, had been known by sailors for more than two centuries, having been named after the same Captain Bligh who achieved a different kind of notoriety as the victim of a mutiny on the

HMS *Bounty*. The accident, coincidentally, took place during the 200th anniversary year of the mutiny.

In many ways, the eras before and after the stroke of midnight on Good Friday, 1989—years of reasonably successful operation, followed by the largest marine oil spill in the history of United States, at least at that time—could scarcely seem more disparate. In another sense, however, perhaps they could not have been more closely related. It is entirely possible that the accident of Good Friday, 1989, would not have occurred but for the tragic complacency that had been engendered by the dozen good years that had passed before. Perhaps, risk analysts have noted, nothing recedes like success. It may have been the very "success" of earlier trips in and out of Prince William Sound, in other words, that helped to make possible a situation in which the captain had retired to his quarters, the ship was under the control of a third mate who was not supposed to be at the helm, the Coast Guard personnel on duty were no longer bothering to monitor even the lower-power radar screens that remained at their disposal after cost-cutting efforts a few years earlier—and at least 11 million gallons of crude oil fouled a thousand miles of once-pristine Alaska shoreline.[7]

The problem, notably, can be expected to operate at an organizational level, not just an individual one. Virtually all organizations, public or private, are likely to face pressures to control costs, and at least one pattern of responses is likely to be consistent—organizations will seek to protect what they consider to be "core" functions, while cutting back on things they consider to be peripheral.

There is a tremendous range of variation in what organizations see as being their core functions—from building cars to busting criminals—but virtually no organization has a truly central focus on increasing the safety of its *own* operations. Instead, protecting health, safety, and the environment tend to be secondary or "while" concerns: Organizations seek to produce energy "while" protecting the environment, operate submarines "while" providing an adequate level of protection for the crew, dispose of wastes "in an environmentally acceptable manner," and so forth. Almost never is risk management included in the first half of the sentence, at least in the description of overall organizational goals, as in "increasing the level of safety for workers and nearby communities 'while' maintaining adequate profit margins"—save perhaps when risk management professionals use such terminology in the attempt to increase their organizations' attentiveness to issues of risk and safety.

In the case of the *Exxon Valdez*, unfortunately, investigations found pervasive evidence that virtually all of the organizations involved—including but by no means limited to Exxon—had been making just such cut-backs in the "secondary" or "peripheral" costs of risk management. One investigation that was published in the *St. Louis Post-Dispatch*, for example, found that "a plan to avert a tanker disaster was developed a decade ago and then gradually dismantled piece by piece." Among other details, the pattern included

• rejection by the Coast Guard . . . of a 1976 state study that forecast tanker accidents and urged such requirements as double-hulled tankers and tug boat escort beyond Bligh Reef. . . .

• two cutbacks in recent years that probably removed an extra pair of eyes that might have spotted the off-course *Valdez*. In 1978, the Coast Guard reduced the distance that local pilots had to guide departing tankers and, since 1984, the Coast Guard has cut its radar staff in Valdez to 36 from 60, reduced the radar wattage and decreased the distance required for radar monitoring. . . .

• Disbandment in 1982 of the Emergency Response Team . . . . Spill-fighting equipment on hand was below the minimum required; even the barge designated to carry containment booms and supplies was in dry dock. . . . [and]

- Carelessness by the state agency charged with [overseeing] compliance. The crash in oil prices in 1986 forced state budget cuts that reduced the work week at the Department of Environmental Conservation to 4 days.[8]

In the case of the *Exxon Valdez*, the atrophy of vigilance also affected other private companies, including the Alyeska Pipeline Service Company—which operated the pipeline and its associated oil terminal in Valdez—as well as the government agencies that were supposed to be keeping an eye on things. When the U.S. Coast Guard shut down its high-power radar in an effort to cut "needless" expenses, for example, that meant that the agency's employees on duty that night—whose numbers had also been reduced in the name of reducing "unneeded" costs—would not have been able to spot the tanker heading toward the rock, even if they had been looking. Even the state of Alaska, which by that time was obtaining 85 percent of its income from the oil industry, showed nothing like the kind of righteous wrath that its officials would start to express once the state began to taste what many wags started calling the "Exxon cocktail"—oil on the rocks.

Still, based on past work in the field of risk analysis, it is not surprising that the company to experience the blowout in the spring of 2010 would have been BP.

Enter BP

The company known today as BP is one that been known for almost fifty years as British Petroleum. Early in the twenty-first century, the company officially shortened its name to its initials, launching as well a massive public relations campaign to proclaim that their company was going "Beyond Petroleum." The company's CEO at the time, John Browne, was an energy visionary; he earned a significant level of good will from environmental activists when his company publicly broke with most of the major firms in the oil industry, acknowledging a possible link between carbon emissions and global warming. BP chose not to join with many other large oil companies in fighting against the ratification of the Kyoto Protocol for controlling global climate disruption, at least in part because BP chose to recognize rather than attack the overwhelming scientific evidence, which shows that carbon emissions are already contributing to significant warming. Browne was even featured in the "green issue" of the magazine *Vanity Fair* in 2006. After Browne's departure from the position in 2007, he was followed by his hand-picked successor, Tony Hayward, and the *Washington Post* made the point of noting that the new CEO had also "committed himself to reform."[9]

For much of the current century, however, critics have called into question the company's actual commitments to reform and its other stated principles. BP, for example, might not be active in *political* fights against efforts to control carbon dioxide and other "greenhouse gases"—so called because, like the glass on a greenhouse roof, they trap heat in the atmosphere, helping to heat up the planet—but at the same time, BP has long been a major investor behind the carbon-heavy and highly controversial efforts to extract petroleum from the huge oil sands reserves of Alberta, Canada. The questions became even more pronounced after Tony Hayward took the reins. He pleased Wall Street by slashing 7,500 jobs, cutting some $4 billion in corporate costs, and boosting profitability, but a number of his cuts raised concerns about whether his only real commitment was to cutting corners, rather than to the pursuit of what most people would consider reform.[10]

Those cuts, moreover came on top of other indicators that the company's commitment to the protection of worker safety and the environment might have been less than stellar, even before Hayward took over. It was BP, for example, that owned the refinery in Texas City, Texas, that exploded in 2005, killing 15 workers and injuring 170 others. It was also BP that owned a six-mile-long section of problematic, corroded pipeline in Alaska that broke open a year later, spewing nearly 200,000 gallons of crude oil across the snow. It was also BP that—while the Macondo spill was still spewing, from April 6 to May 16, 2010—burned 500,000 pounds of toxic chemicals at that same Texas City refinery, not bothering to notify residents until weeks later that the burning had released 17,000 pounds of cancer-causing benzene.[11]

In the case of the 2006 Alaska spill, subsequent investigations showed that the company had been warned to check the pipeline four years earlier, but had failed to do so. The company was found guilty of a misdemeanor violation of the federal Water Pollution Control Act, being fined $12 million. A congressional committee concluded that BP had ignored opportunities to prevent the spill and that—just as the literature on the atrophy of vigilance would predict—the spill reflected "draconian" cost-saving measures and safety shortcuts.[12]

In the case of the 2005 refinery explosion, the company pleaded guilty to federal felony charges, and the Occupational Safety and Health Administration (OSHA) levied the largest fine in its history at the time—$21 million. Investigators later determined that the company had ignored its own protocols on operating a refinery tower that was filled with gasoline, and—in an ominous precedent—that a key warning system had been disabled. By October 2009, OSHA fined BP again, this time for the company's failure to make the safety upgrades it had already agreed to make under the settlement agreement. After a delay of some ten months, BP achieved "the dubious distinction of topping the previous record OSHA fine of $21 million," agreeing to pay what became the new, all-time record fine for OSHA violations—$50.6 million. Still, those fines were tiny in comparison with the company's profits, and in last announcement of corporate quarterly profits before the Macondo blowout, Hayward announced that the company still saw "significant opportunity" for more cost-cutting, all as part of his ongoing efforts to improve the company's profitability.[13]

Strictly speaking, it appears that, particularly in the case of refinery safety, BP's actions ran counter to the predictions from work on the atrophy of vigilance. According to the theory, BP should have shown dramatically increased vigilance, at least in the first few years after the accident. Instead, a study of records from the Occupational Safety and Health Administration (OSHA) later pointed to BP as being the "Renegade Refiner." In the months leading up to the *Deepwater Horizon* blowout, between mid-2007 and early 2010, BP single-handedly accounted for nearly half of all OSHA safety citations to the entire refining industry. No other oil company even came close: BP had 862 citations, while second-place Sunoco got 127, and third-place ConocoPhillips had 118. The record was even more lopsided for the worst offenses. BP received 69 citations for "willful" violations, defined as those including "intentional disregard for employee safety and health"—triple the number for the all of the rest of companies in the refining industry, in combination. Most spectacularly, though, BP received 760 citations—out of a grand total of 761 for the entire industry—for "egregious willful" violations, or the worst violations of all, reflecting "willful and flagrant" violations of health and safety laws.[14]

These were not good signs.

Internal company documents obtained by investigative reporters for ProPublica and the *Washington Post* suggests that these problems may indeed have been part of a broader pattern:

A 2001 report noted that BP had neglected key equipment needed for an emergency shutdown, including safety shutoff valves and gas and fire detectors similar to those that could have helped prevent the fire and explosion on the Deepwater Horizon rig in the Gulf.

A 2004 inquiry found a pattern of the company intimidating workers who raised safety or environmental concerns. It said managers shaved maintenance costs by using aging equipment for as long as possible. Accidents resulted, including the 200,000-gallon Prudhoe Bay pipeline spill in 2006—the largest spill on Alaska's North Slope—which was blamed on a corroded pipeline.

Similar problems surfaced at BP facilities in California and Texas....

BP has had more high-profile accidents than any other company in recent years....

To avoid having its Alaska division debarred—the official term for a contract cancellation with the federal government—the firm agreed to a five-year probationary plan with the EPA. BP would reorganize its environmental management, establish protections for employees who speak out about safety issues, and change its approach to risk and regulatory compliance.

Less than a year later, employees complained to an independent arbitrator that the company was letting equipment and critical safety systems languish at its Greater Prudhoe Bay drilling field. BP hired independent experts to investigate.[15]

After the *Deepwater Horizon* exploded, more detailed investigations by the Coast Guard and Congress revealed further evidence of problems. An Associated Press report on Coast Guard investigations quoted Truitt Crawford, a roustabout for drilling rig owner Transocean, as saying he "overheard upper management talking saying that BP was taking shortcuts by displacing the well with saltwater instead of mud without sealing the well with cement plugs, this is why it blew out." Other worker statements and a congressional memo about a BP internal investigation of the blast provided further evidence that warning signs were often being ignored. According to the congressional memo, even tests that were done less than an hour before the well blew out "found a buildup of pressure that was an 'indicator of a very large abnormality,' BP's investigator said."[16]

In the lingo of risk assessment, such patterns of safety shortcuts often indicate the presence of "precursors" to more serious disasters. Even in simpler, everyday language, these are not the kinds of patterns that would normally inspire an outburst of increased confidence. Based on the risk assessment literature, though, there would be reasons to expect that such corner-cutting patterns would have been relatively widespread within BP, and that does in fact appear to have been the case.

In the aftermath of the blowout in the Gulf, for example, a report by CNN also found that BP had been "trying to shut down" the internal safety watchdog office it had set up under congressional pressures after the Texas City refinery explosion. "Sources close to the office," according to the investigation, said "BP doesn't like having independent investigators pursuing those complaints. A union representative told CNN that some workers who complained had faced retaliation." The story went on to note that 112 employees had filed complaints with the office by that time, with 35 of the complaints having to do with "system integrity or safety issues" that the office described as "extremely serious." The report even quoted one source inside the internal safety watchdog office, who said simply, "I'm surprised we're still here."[17]

The problems were sufficiently widespread that they can be illustrated with the information that came to light in just one day's hearings—those being held jointly by the Coast Guard and the Interior Department on July 20, exactly three months after the blowout. First, Ronald Sepulvado, the company's well manager, testified that he had notified supervisors in Houston about a leak from the blowout preventer, doing so just days before the spill. In that same day's hearing, investigators focused on a report that was prepared on April 18, two days before the blowout, by the company that was under contract to cement the well casings into place and plug the well, namely Halliburton. The report predicted that the then-current BP well design could risk "severe" gas flow problems. That same day's hearings also focused on two BP maintenance audits from February and March, indicating "a number of mechanical problems on the rig, including an engine that was out of operation, a thruster that was not running and a leak in the blowout preventer."[18]

That might seem to be unsettling enough, but in hearings that were held three days later, a chief electronics technician for Transocean, Mike Williams, testified that the disabling of a key warning device in BP's Texas City refinery might not have been an isolated occurrence. Instead, he testified that a number of key safety systems on the *Deepwater Horizon* had been bypassed or disabled—some for months or years. He even reported that he had been "chewed out" after reactivating a gas safety valve that had been placed in bypass mode. He recalled a Transocean subsea supervisor, Mark Hay, as saying, "The damn thing has been in bypass for five years. Why did you even mess with it? ... As a matter of fact, the entire fleet runs them in bypass."

The explanation that Williams got for the bypassing the alarm, incidentally, was that the rig's managers did not want workers to wake up at 3:00 a.m. because of false alarms. One wonders if they might have felt differently about "true" alarms. The valve, which was designed to cut off natural gas flow if pressures got to be too high, was in bypass mode when natural gas from the oil well shot up into the rig and exploded.[19]

To repeat, however, it would be a mistake to focus on BP alone.

Transocean, for example—the company from which BP was leasing the *Deepwater Horizon*—also had a pattern of near misses, although at least Transocean responded by engaging the risk-management company, Lloyd's Register, to investigate its Gulf of Mexico operations. According to internal documents obtained by the *New York Times*, that investigation identified "critical equipment items that may lead to loss of life, serious injury or environmental damage as a result of inadequate use and/or failure of equipment." The *Times* investigation also shed potential light on why the Deepwater Horizon sank, noting that the rig had long had problems with its ballast system and had needed to evacuate more than 70 employees in May of 2008 because the rig was listing to such an extent that part of it flooded.[20]

The *Washington Post* took note of a somewhat broader pattern: Even federal government "investigations" have often relied on reports from private contractors that were hired by the very firms being investigated, or by the firms themselves. In the same month that the *Deepwater Horizon* exploded, Transocean gave parts of two cranes its worst rating, indicating that they did not work or should be removed from service, but "That assessment was for the company's internal use. Less than two months earlier, one of the main inspection firms upon which governments depend declared that the same cranes were in satisfactory condition." Four months after the blowout began, the Wall-Street rating firm, Moody's Investors Service, downgraded its rating for Transocean to Baa3—just above "junk bond" status—and added a "negative outlook."[21]

The most systematic expression of concerns from the early investigations, however, focused largely on BP itself. Perhaps the most concise summary of the concerns was the one provided in a 14-page letter that the chairs of the House Committee on Energy and Commerce and of its Subcommittee on Oversight and Investigations (Henry Waxman and Bart Stupak, respectively) sent to BP's CEO, Tony Hayward. The letter was detailed and specific, but again, its overall message pointed toward an overall pattern of safety shortcuts that appeared to have contributed to the *Deepwater Horizon* disaster. As the letter noted, BP and its partners also gave evidence of having downplayed or ignored warning signs that the well in question was one for which corner-cutting would have been particularly ill-advised. On April 15, for example, five days before the explosion, BP's drilling engineer had called the operation a "nightmare well." The drilling operators had experienced a series of kicks—those powerful spurts of natural gas, under high pressure—providing ominous warnings of the kind of blowout that actually took place on April 20. The letter also noted that—despite the well's difficulties, and perhaps in part because the drilling operation was behind schedule —

> BP appears to have made multiple decisions for economic reasons that increased the danger of a catastrophic well failure. In several instances, these decisions appear to violate industry guidelines and were made despite warnings from BP's own personnel and its contractors. In effect, it appears that BP repeatedly chose risky procedures in order to reduce costs and save time and made minimal efforts to contain the added risk.[22]

These and other investigations will continue for years, meaning that it is far too soon to reach any final verdicts. Even based on the limited information that has already come out, however, it does appear that BP and its partners made a series of fateful decisions, each of which increased risk, and almost all of which appear to have been designed to save time and money. In light of the evidence that had come to light by the time of the Congressional letter, four of those decisions seem to have been particularly noteworthy.

The first was the way in which BP chose to install the final section of casing—the lowest part of the collapsible telescope noted above. BP decided not to attach this last and longest segment to the bottom of the next-to-last segment, but instead, to run this casing straight through from the top of the well to the bottom. This removed one of the seals, and it meant that the integrity of the well depended entirely on the cement that held the last casing in place. This option saved both time and money, but it was not the most prudent choice for a well that had been having so many problems with high-pressure kicks. This particular shortcut was not even included as an option in BP's original application to Minerals Management Service, so the choice required an application for an amended permit. Eager to help, though, the agency approved the amendment that very same day.

The second decision involved "cementing in" the final casing. In this process, it is important that the annular space be uniform—that is, that the casing be a uniform distance from the rock formation around it—so that cement can flow into the space evenly, making a proper bond. The usual way to assure even spacing is to use *centralizers*—steel, often cylindrical spacers, which are placed around the casing to hold it in the center of the well that has just been bored. "Halliburton, the contractor hired by BP to cement the well, warned BP that the well could have a 'SEVERE gas flow problem' if BP lowered the final string of casing with only six centralizers instead of the 21 recommended by Halliburton." BP, however, ignored this advice, proceeding with just the six centralizers that were on board, rather than waiting for 15 more to be located and brought to the rig. That saved the company time and money, at least in the short term, but at the potential risk of affecting the integrity of the cement job.[23]

The third decision was also related to the cementing of the final casing. The standard practice is to circulate the entire column of mud, from top to bottom, to remove any pockets of gas and also to remove debris that might otherwise compromise the cementing. In fact, BP's drilling plan called for the mud to be circulated one and a half times. Unfortunately, that procedure would have taken about 12 hours longer than the short-cut BP decided to use—circulating only a portion of the mud—and BP decided to bypass the usual practice in this case, as well.

Fourth and finally, a test called a *cement bond log* is supposed to be run after the casing is cemented in, to test its integrity. This is an acoustical test that can locate voids in the bonding of the casing to the formation. A crew from Schlumberger, an oil-field contractor, was on board the rig on the morning of April 20th to run such a test. The cement bond log would have taken 9 to 12 hours and cost over $128,000. Canceling the test cost $10,000 and no down time—at least that morning. The crew from Schlumberger was told their services were not needed. They left the *Deepwater Horizon* at 11:15 a.m. on April 20th, about ten and a half hours before the blowout.

It is important to understand that these decisions were cumulative—they all added to the ultimate risk of a blowout. BP might have been able to get away with shortcuts in the casing design, for example, if the company had used the recommended 21 centralizers and circulated the mud to remove obstructions to the even distribution of cement. They might well have been able to get away with any two of these shortcuts if they had been more vigilant in other ways—being alert to other warning signs of problem that needed to be fixed, or for that matter, being careful *not* to put gas safety valves in bypass mode. Given that deepwater drilling pushes the safety limits of present-day technology, most companies do not take such shortcuts most of the time. If they did, we probably would have had other catastrophic Macondo-like blowouts by now.

Years ago, a computer back-up system was advertised with the slogan, "Blessed are the Pessimists, for they hath made Backups." By the time of the blowout on April 20, 2010, by contrast, BP had remarkably few remaining backups—or to quote Congressional investigators, the company had "a well design with few barriers to gas flow." The corner-cutting was particularly unfortunate in that—as anyone familiar with petroleum geology should have been able to predict, and as the blowout proved—the company was drilling in a deep-water location where there was good reason to be worried about high gas pressure. Unfortunately, the investigators noted, "the common feature" of the decisions that BP made in the face of such pressures was that they posed "a trade-off between cost and well safety."[24]

The decision to remove the drilling mud removed yet another "barrier to gas flow." Even if, as indicated by the evidence available to date, the final cementing of the casing was faulty, appropriate use of drilling mud could have reduced the risk. As discussed earlier, the virtues of drilling mud include the fact that it is thick and heavy. Unfortunately, in the case of the mud, too, initial investigations point to a pattern of corner-cutting that, in the end, proved tragic. BP chose to remove the drilling mud, believing—erroneously, it appears—that the cementing of the casing was secure:

> Testifying before a Coast Guard and MMS board ... Jimmy Harrell, the Deepwater Horizon's offshore installation manager, said BP initially wanted to replace heavy drilling lubricant ... without performing a negative-pressure test. He said the plan to proceed with removing the drilling mud came from BP's Houston headquarters and had not been approved by MMS.
>
> Harrell refused, however, to go forward without the negative-pressure test, which involves sucking the air out of the pipeline to see if gas or oil leaks into the pipe. The test is critical to be certain that the pipe is secure before the heavy mud, which is the primary force keeping gas from surging up the pipe, is removed....
>
> Two negative-pressure tests were performed and they indicated leaks, Harrel said. On the first test, 23 barrels of drilling mud flowed out onto the Deepwater Horizon's drilling floor. A second test returned

15 barrels, according to Harrell. Ideally, a negative-pressure test returns no drilling mud.[25]

Within the field of risk analysis, patterns of short-cuts such as these would be expected to set off of a series of flashing red warning lights, at least as vivid as the one that failed to stop the *Exxon Valdez* from slamming into Bligh Reef. The reasons are suggested by the headline of an on-line story from CNN. Money.com: "BP Was Warned: Interviews with Employees and a 2002 Letter Predicting 'Catastrophe' Show That BP's Problems Should Have Come as No Surprise to Management."[26]

Yet the problems also should have come to no surprise to the federal agency that bears responsibility for regulating offshore oil operations—the one that gave same-day approval to BP's request to skip the previously planned step of attaching the last section of casing to the next-to-last segment, namely the U.S. Minerals Management Service (MMS). The MMS, which is part of the U.S. Department of Interior, was originally established under the Reagan administration, by a man whose proclivities were so infamous in environmental circles that he inspired the collection of political cartoons named *100 Watts: The James Watt Memorial Cartoon Collection*. That, however, was just the beginning.[27]

Sex Is not Arm's Length

Over the years, MMS has often been criticized by environmental activists as being a toothless watchdog, but a 2008 Inspector General report concluded that the agency's relationships with oil companies were even worse than most critics had claimed. It was a matter of being in bed with the industry, literally, in a pattern of sex, drugs, and the wrong kind of role. The investigation discovered what the report called "a culture of substance abuse and promiscuity" in which employees accepted gratuities "with prodigious frequency." More than a dozen employees, including the former director of the oil royalty program, took meals, ski trips, sports tickets, and golf outings from industry representatives. Officials accepted gifts, engaged in drug use, and yes, even had sex with employees of the energy firms from which they were expected to collect royalties. In case there might have been any confusion about why that was wrong, the Inspector General explained, "Sexual relationships with prohibited sources cannot by definition be arms-length." In the words of Philip Verleger, a former Treasury Department employee, now with the University of Calgary, who was quoted in a report on National Public Radio, "When I was at Treasury, you would've gone to jail if you'd done any of those things. Yeah, just read that Inspector General's report, just a little bit of it. There was no gray line. I mean, these guys, they were way over it."[28]

The sex and drugs were connected to the agency's royalty collection division, which adds new emphasis to the old adage "follow the money," but in the aftermath of the *Deepwater Horizon* disaster, reporters learned more as they looked into the safety record of the agency with a bit more intensity. One of the first things they noticed was that, in 2009, the MMS gave its regional Safety Award for Excellence (SAFE) award to Transocean—the company that owned the *Deepwater Horizon*, and that Lloyd's Register had found to have "critical equipment items that may lead to loss of life, serious injury or environmental damage." The Minerals Management Service evidently saw things differently; it offered its annual award to recognize what the agency considered "outstanding drilling operations" and a "perfect performance period." BP had been named as a finalist in the 2009 competition, and for the 2010 Awards luncheon—scheduled to take place less than two weeks after the sinking of the *Deepwater Horizon*—BP was a contender for two awards, both relating to the safety of the company's offshore oil work.

In a decision that was attributed to the need to focus on "the ongoing situation with the Transocean Deepwater Horizon drilling accident"—rather than to any inconvenience in its timing—MMS canceled the May 3 ceremony, announcing it would reschedule for a later date.[29]

The MMS, unfortunately, was also the agency that approved BP's Oil Spill Response Plan—a 582-page document designed to cover all BP operations in the Gulf of Mexico. It was a spectacular illustration of what the noted scholar Lee Clarke has called a "fantasy document."[30]

In the plan, which was submitted under President Bush, although it received its final agency approved a few weeks after President Obama took office, BP actually calculated a worst-case scenario:

> BP has determined that its worst case scenario for discharge from a mobile drilling rig operation would occur from the Mississippi Canyon 462 lease [about 28 miles southwest of MC 252]. MC 462 is a planned exploration well targeted for Miocene oil reservoirs. Given the anticipated reservoir thickness and historical productivity index for the Miocene, worst case discharge is expected to be 250,000 barrels of crude oil per day.[31]

In the same document, however, BP comfortingly claims that to have had at its disposal an amazing collection of skimming equipment, with a "recovery rate of 491,721 barrels/day." Had that been true, BP would have been able to remove an incredible 20,652,282 gallons, or roughly twice the total volume of the *Exxon Valdez* spill, every day—or to put things differently, more than eight times the estimated spill volume of 60,000 barrels per day that the company actually had so little success in controlling. Still, because the well was being drilled in the central Gulf of Mexico, section 7.1 of the MMS-approved plan decreed that "a site-specific Oil Spill Response Plan (OSRP) is not required."[32]

It appears the agency decided that not much fact-checking was required, either. The go-to wildlife expert listed in the plan, Professor Peter Lutz of the University of Miami, had left that institution twenty years earlier, and in a particularly inconvenient detail, he had died four years before the plan was approved. The plan claimed there would be only a 21 percent chance that oil from a spill would reach the Louisiana coast within a month—while the actual spill, with a probability of 100 percent, took only nine days to start fouling the coastline. The plan claimed there would be "no adverse impact" for sea turtles or endangered marine mammals, both of which started showing up dead within a few days of the blowout. The plan did, however, express concern for walruses, sea otters, sea lions, and seals, including all of them under the heading "sensitive biological resources"—even though not a single one of them has lived in the Gulf for the last several million years. Perhaps, noted one wag, we should not place so much emphasis on the fact that the plan fails even to mention an environmental feature as important as the Loop Current—which comes to within a few dozen miles of the spill before it wraps around the south tip of Florida and heads up the Atlantic coast. If the Loop Current were to carry the oil spill far enough, it could eventually head toward Greenland, so maybe BP officials were thinking that nearby currents would carry their spill all the way up to the part of the world where walruses and sea lions might need protection.[33]

Maybe. Another possibility, though, was expressed by Ian Masters, the host of a public radio show in Los Angeles, who interviewed the two of us a few days after the disaster. After noting that MMS was a creation of James Watt, whose scorn for environmentalists was legendary, he suggested that putting Watt's agency in charge of protecting the environment from oil spills might be "like putting Count Dracula in charge of the blood bank."

In the aftermath of the spill, MMS officials expressed a distinctly different view, insisting that they conducted safety investigations almost continuously. "In an e-mail to AP, an Interior Department official emphasized with italics that the MMS inspects rigs 'at least once a month' when drilling is under way." Similarly, at a joint Coast Guard–MMS investigatory hearing in Kenner, La., Michael Saucier, MMS's regional supervisor for field operations in the Gulf, was asked how MMS performs drilling inspections in the Gulf of Mexico. "'We perform them at a minimum once a month, but we can do more if need be,' Saucier said."[34]

Such claims, however, are difficult to square with the MMS's own records, which showed that "Federal inspectors failed to conduct nearly a third of required inspections on the Deepwater Horizon rig in the 28 months before it exploded and sank in the Gulf of Mexico." The relevant MMS reports, including the last one, dated just three weeks before the April 20 explosion, "indicate[d] that the rig's blowout preventer was functioning properly, and they make no mention of any persistent problems with surges of natural gas, or 'kicks,' flowing up through the well and disrupting drilling.... MMS inspectors noted the presence of a kick in October 2008, but none later."[35]

As the risk analysis literature would predict, though, the problem was scarcely limited to the Macondo well. The *Washington Post* reported that a total of 12,087 oil-related incidents in the Gulf had been reported to the Minerals Management Service during the five years before the Macondo blowout—including one incident on a neighboring rig, leased by Louisiana Land Oil and Gas, a year and one day earlier. Crew members on that rig "reported hearing a 'deafening roar' as fluids shot up, knocking over huge metal equipment on the deck." The good news, relatively speaking, is that the earlier incident remained "just" a near-disaster. The bad news is that not even such a close call was enough to provide the kind of wake-up call that was needed. Instead, in the words of the *Post* article, "The number of accidents, spills and deaths regularly occurring in the region has far surpassed the agency's ability to investigate them."[36]

That was certainly the pattern in the case of the *Deepwater Horizon*. As summarized in an Associated Press report,

> Earlier AP investigations have shown that the doomed rig was allowed to operate without safety documentation required by MMS regulations for the exact disaster scenario that occurred; that the cutoff valve which failed has repeatedly broken down at other wells in the years since regulators weakened testing requirements; and that regulation is so lax that some key safety aspects on rigs are decided almost entirely by the companies doing the work.... In response to a Freedom of Information Act request filed by AP, the agency has released copies of only three inspection reports—those conducted in January, February and April [of 2010]. According to the documents, inspectors spent two hours or less each time they visited the massive rig. Some information appeared to be "whited out," without explanation.[37]

From 2001 to early 2009, of course, MMS reported to an administration that was headed by two Texas oil men, George W. Bush and Richard Cheney, both of whom were famously hostile toward federal regulations, preferring to trust the Magic of the Marketplace and pushing for an "Energy Plan" that relied heavily on secret input from some of the nation's biggest oil companies. The more environmentally oriented Obama administration that came into office in 2009, however, had made relatively few changes in the agency that would have affected its day-to-day operations, at least before the spill. Instead, the new administration's most visible change to the agency came soon after the spill, with a decision to divide the royalty-collection and enforcement functions of the agency, as well as coming up with the new name of "Bureau of Ocean Energy Management Regulation and Enforcement"—variously abbreviated as BOEM, BOE, or BOEMRE.[38]

Still, even after the spill, as court battles waged over the Obama administration's proposal to impose a six-month moratorium on all deep-water drilling in the Gulf of Mexico, an investigation by McClatchy newspapers found little evidence that the Obama administration had been taken over by heavy-handed regulators. Instead, MMS continued to issue exemptions—"categorical exclusions," in agency-speak—at a rate of more than one per day, "despite President Obama's vow that his administration would launch a 'relentless response effort' to stop the leak and prevent more damage to the Gulf." The waivers included full exemptions from requirements for detailed environmental studies for a BP's plans for exploratory drilling at a depth of more than 4,000 feet, as well as an exploration plan by Anadarko, one of the partners on the Macondo well, at more than 9,000 feet.[39]

Congress, of course, is supposed to keep watch over the agency watchdogs when the White House fails to do so, but Congress also showed little interest in the safety of oil drilling operations, at least until after the spill. The specific members of Congress who are beneficiaries of industry prosperity would almost certainly deny any connections between their personal fortunes and their voting patterns, but the *Washington Post* did note that nearly 30 members of the Congressional committees charged with keeping an eye on oil and gas companies held millions of dollars of investments in the industry. "In 2008, members of the five committees held investments worth at least $8.1 million in companies they oversaw"—an amount that grew by another million dollars or so by the end of 2009. The lawmakers also held substantial investments in the specific companies involved in the *Deepwater Horizon*. As the article noted, "At the top end of the estimates, lawmakers on those panels may have held just shy of $1 million in shares of BP; Transocean, which owned the oil rig; and Anadarko Petroleum, a lease partner."[40]

Representative Joe Barton—the senior Republican on the House Energy and Commerce Committee, who happens to hail from Texas—evidently had particularly strong feelings about the injustice of the anger that was directed toward BP after the spill. When BP's CEO, Tony Hayward, appeared before that Committee, Barton said that what he considered a "tragedy" was not so much the spill itself, but the fact that a "a private corporation can be subjected to what I would characterize as a shakedown" by the Obama administration.

After a delay of several hours, the Congressman did apologize for his remark, but a report for the *Dallas Morning News* noted that the statement was not much of an aberration from the industry-friendly positions that Barton had staked out in his 25 years in Congress. As campaign finance documents make clear, though, it is not just that Representative Barton, who owns a gas well of his own, likes the petroleum industry. The petroleum industry likes him as well. His top contributor during the dozen years leading up to the BP blowout was BP's partner on the well in question, Andarko Petroleum, which single-handedly poured nearly $150,000 into the Congressman's campaign coffers. The oil and gas industry as a whole donated ten times that amount to him—$1.5 million—over the two decades leading up to the spill. For comparison purposes, Barton's Democratic opponent in the 2010 election, David Cozad, had the princely sum of $253 in cash on hand in his own campaign war chest at the time of the story. As the *Dallas Morning News* reported, Barton's Congressional seat was generally "considered safe."[41]

Even the third branch of American government, the judiciary, may provide only an imperfect set of safeguards. The U.S. District Judge who overturned the initial proposal from the Obama administration for a six-month moratorium on deepwater drilling, Martin Feldman, was found by an Associated Press investigation to own stock in Transocean, the company that owned the *Deepwater Horizon*. His 2008 financial disclosure report—the most recent one available at the time of the decision—also showed investments in Halliburton, the cementing contractor on the *Deepwater Horizon* job, and in a series of other companies that were affected by his decision, including Ocean Energy, Prospect Energy, Peabody Energy, Pengrowth Energy Trust, Atlas Energy Resources, Parker Drilling, and others.[42]

Still, Judge Feldman might not be that unusual. Instead, at least according to financial disclosure reports from 2008—again, the most recent ones available at the time of the blowout—the problem included, literally, the majority of the judges in the region. "Thirty-seven of the 64 active or senior judges in key Gulf Coast districts in Louisiana, Texas, Alabama, Mississippi and Florida have links to oil, gas and related energy industries, including some who own stocks or bonds in BP PLC, Halliburton or Transocean—and others who regularly list receiving royalties from oil and gas production wells."[43]

Finally, these patterns do not operate in isolation from one another. There are numerous connections between the industry and all three branches of the government—monetary and otherwise. The result is a system in which people move back and forth among the available positions, and the distinctions between regulators and the regulated sometimes get so blurred that they disappear. The *Washington Post* has noted that, even in comparison to "the usual revolving-door standards on Capitol Hill," the oil industry proves to be notable, with three out of every four of the industry's Washington lobbyists having previously worked in one capacity or another for the federal government. In such a system, predictably, a common world view is likely to emerge—one that favors certain interests and slights others. The interests that are most likely to get left out of that system, unfortunately, are those of the broader public.[44]

All in all, the pattern that is visible in the case of the *Deepwater Horizon* is one that carries echoes of a question from a respected academic article that was published almost a quarter of a century earlier, just before the *Exxon Valdez* spill: Who will watch the watchers? The question still doesn't have a very comforting answer.[45]

According to at least some economists with whom we have discussed this disaster, economic self-interest alone should have been sufficient to motivate BP to minimize the risks of drilling, and in light of estimates of the company's liabilities from the spill, which run into tens of billions of dollars, the company's officials may well wish today that they had not worked quite so hard to speed up the drilling, saving the comparatively small drilling costs of half a million dollars a day. "Free-market incentives," though, clearly did not prove sufficient to maintain BP's commitment to safe operations in advance of the blowout.

At the same time, while BP is at the center of the mess, it was not the only organization involved in the disaster, and it should not be the sole focus of scrutiny. Even before the flow of oil was finally shut off, it became clear that the precautions of the past were woefully inadequate. Informal discussions with friends in the drilling industry indicate, incidentally, that such a judgment should not come as a surprise. Instead, as one of them put it to us, people in many other oil companies were thinking some variation on the theme of, "there, but for the grace of God, go I."

Even state and local government officials, many of whom have been seen in the aftermath of the blowout to be irate—positively irate—tended before the spill to be significantly more positive toward the industry. Like Congressman Barton, many of the same state and local officials have long treated even minimal regulatory efforts as being outrageous and unnecessary intrusions into the operations of trusted and valued friends.

Not to put too fine of a point on it, the whole system is flawed. The patterns that have emerged are particularly worrisome in light of the fact that drilling has increasingly been moving out of relatively shallow water depths—where the safety record of drilling for oil, at least in U.S. waters, has indeed been reasonably good—into much deeper waters, literally as well as figuratively. In those deeper waters, as the BP blowout suggests, there are good reasons why we would all want drilling operations to be prepared for new dangers and new levels of challenges. The disaster demonstrated that neither the oil companies nor the relevant government agencies were as prepared as most Americans would have expected—and with problems being that pervasive, a clear possibility is that the roots of the problem may go even deeper, beyond the oil companies and the Minerals Management Service alone.

4 Colonel of an Industry

The founders of the United States had no way of knowing it, but their new nation was born with a treasure chest in the basement. Below the rich natural resources that were visible on the surface, the territories that would ultimately be included in these United States were the home to some of the world's richest deposits of one of the world's most valuable resources—petroleum. It just took the nation's founders, and their descendants, some time to figure that out.

For thousands of years before the descendants of Europeans came to North America, the native peoples on the continent knew about the presence of petroleum, mainly in the form of seeps that brought underground deposits to the surface. The oil had a number of uses, but perhaps the only one that involved a significant contribution to transportation was when native peoples such as the Chumash, along what is now California's south-central coast, used the tarry "pitch" from natural seeps to seal and hence to improve the seaworthiness of their canoes. Up through roughly the middle of the nineteenth century, the European settlers and their descendants made even less use of the substance we know today as oil; instead, when they spoke of "oil," the reference was likely to mean *whale* oil. Given that the electric light had not yet been invented at the time, they were most often thinking about using the stuff in whale oil lamps.

"Yankee" or American whaling started almost as soon as colonists started to settle in North America, which is to say during the 1600s, but it was not until the 1800s when the United States became a virtual OPEC of the whale oil world. By 1833, there were 392 American whaling vessels. That number almost doubled in the next 13 years, reaching 736 whaling vessels by just about the middle of the century, in 1846. By that time, the industry employed some 70,000 people, and U.S. ships made up 80 percent of the entire world's whaling fleet. Herman Melville, soon to be famous as the author of *Moby Dick*, set out on his own whaling adventure from New Bedford, Mass., in 1841, but he had a great deal of company. As noted by the National Park Service, "more whaling voyages sailed from New Bedford" during the 1850s "than from all of the world's [other] ports combined." By the middle of the 19th century, whaling produced roughly 4 to 5 million gallons of sperm oil, plus at least a million pounds of bone, each year.[1]

During the first half of the nineteenth century, very few of the whaling captains would have had reasons to imagine that they would be facing any significant competition from "rock oil," which, perhaps prophetically, was also often called "American oil." Even when the descendants of the early European colonists began to develop uses for that rock oil in the mid-1800s, the emphasis was on its supposed medicinal value. Drawing small quantities of oil from natural seeps, a growing band of businessmen began to tout the product as the cure for almost any illnesses that could be imagined. One of the more imaginative entrepreneurs was a fellow named Samuel Kier, based in Pittsburgh, who was bottling, selling, and extolling the spectacular medical virtues of his product by the late 1840s. He sold almost a quarter of a million bottles of his oil within the next decade, at the price of a dollar per bottle, but the actual curative properties of the contents fell a bit short of the sales pitches. Given that this oil was named after the Seneca Indians, who had inhabited the region of Pennsylvania where much of the oil was gathered, perhaps longest-term contribution from Kiel and his colleagues was linguistic: Their product—Seneca Oil—is remembered today under the slightly altered pronunciation of "snake oil."[2]

Other entrepreneurs, however, were soon to develop other uses for rock oil, thanks in part to forms of inventiveness that, rather than celebrating the oil's claimed medicinal miracles, in its original form, actively worked to change the physical properties of the oil. In 1849, a Canadian named Abraham Gesner developed and soon patented a process for distilling the rock oil to produce *keroselain*—a word derived from combining the Greek words for "wax" and "oil," soon to be shortened to *kerosene*—which provided for the first time a reasonable substitute for the whale oil in all those whale oil lamps. By that time, even Kier, the seller of the snake oil, was looking into other commercial uses for his product. After experimenting a bit, he built what is generally believed to be the first commercial oil refinery, in Pittsburgh, in 1854—a date that is as good as any other for marking the start of America's first full century of oil.

It was not exactly a spectacular start. Kier's first refinery had just a five-barrel capacity, and it was chased out of its downtown waterfront location and moved to the suburbs because his neighbors were worried about the potential for explosions. They had fairly good grounds for their concern. The Yale University chemist, Benjamin Silliman, had done what were probably the era's most important analyses of seep oil, but he had also blown up the equipment in his Yale lab in the process. Still, Kier persevered, and by 1858, he and a local Pittsburgh firm were selling New York City distributors the refined product from his rock oil as an illuminant, or a replacement for whale oil.[3]

Demand for the product rose, and petroleum was soon also finding a use as a lubricant in textile mills. Still, since supply was limited to what could be obtained from natural seeps—and since the prices of whale oil had grown ever-steeper, as the heavy whaling pressure brought several whale species uncomfortably close to extermination—the time was ripe for a technique that could provide more oil from sources other than whales.[4]

Digging Deeper

According to the timeline put together by ASTM International, originally known as the American Society for Testing and Materials, the idea of digging for oil was probably started by builders along the banks of the Euphrates River, in present-day Iraq, who were digging asphalt from oil seeps about 6,000 years ago, using it as mortar between building stones. By 347 AD, oil wells up to 800 feet deep were being drilled in China, using bits attached to bamboo poles. The ASTM site also lists the first modern oil well as having been drilled in Asia, in 1848, on the Aspheron Peninsula, northeast of Baku. The well that started the massive oil industry of today, however, is one that started in Pennsylvania, several years later.[5]

That effort had its origins in 1855, after several speculators came in contact with a bottle of rock oil. Judging that it had commercial potential, they purchased some land near Titusville, Pennsylvania, that had several crude oil seeps. They soon formed the Pennsylvania Rock Oil Co., the first true petroleum company—later to be renamed the Seneca Oil Company. In the winter of 1857, to investigate the possibility of exploiting the seeps on their newly purchased land, the investors sent to

Titusville a vagrant with no technical or mining expertise, but with a mythical title of "Colonel," namely Edwin L. Drake. According to at least one account, Drake "happened to be out of work and in the same hotel in New Haven, Connecticut, where the founders of the new Seneca Oil Company were staying.... Drake got the assignment because he had a free pass on the railway due to his former jobs as express agent and conductor."[6]

Drake made a reconnaissance visit in late 1857 and then came back to Titusville in May of 1858, where he tried to increase the flow of the oil seeps. After a number of unimpressive efforts that succeeded only in increasing the flow up to about 10 gallons a day, he decided to try out some of the same techniques that had previously been used to drill salt wells in the area.

By today's standards, Drake's well was almost hopelessly modest, involving a wooden frame and a drill bit that was initially powered by human labor. Still, Drake also made an important innovation that allowed the drilling to get past the layer of gravel that initially kept him from getting to the underlying rock, namely a cast-iron *drive pipe*—the forerunner of the casings that would later be used in oil wells around the world. Later historians would ultimately see this small well as the start of the most powerful industry the world had ever known, but his drilling effort was ridiculed as "Drake's Folly" by local observers at the time, "because of its apparent worthlessness."[7]

In the end, the margin between folly and destiny may have been formed by two remarkable strokes of good luck. The first was a matter of timing. The financial support that Drake received from the Seneca Oil Company appears not to have been much greater than the level of moral support he received from his neighbors. Drake ran out of company money, then ran out of his own money, and finally borrowed from friends and from a bank in nearby Meadville, Pennsylvania. Despite his dedication, however, he had still not managed to hit oil by late August, and his backers sent him a note telling him to call the effort to a halt. The letter, remarkably, arrived on Monday, August 29, 1859. At the end of the drilling on Saturday afternoon, two days earlier, the drill bit slipped six inches past the bottom of the 69-foot-deep hole that he and his driller, Billy Smith, had managed to dig. On the next day, Sunday, August 28—literally the day after they quit and the day before they received the letter telling them to do so—Billy Smith and his young son Samuel looked down into the hole and saw that it was filling up with crude oil.

The second stroke of luck involved the local geology: Subsequent analyses have concluded that, if Drake had moved his initial drill rig just a few yards in almost any direction, he would still have been a hundred feet or more away from the oil. As things turned out, however, by the time that his backers' letter had arrived in Titusville, so had the birth of what Sampson was later to call "the world's biggest and most critical industry."[8]

Drake's well—with its depth of just about 70 feet, and a production of just 25 barrels per day, even after a pump doubled his output—also helped to bring on the world's first "bust" in petroleum prices. Fifteen months after Drake's well "came in," so had some 75 other wells nearby, and the drillers began to learn first-hand about the laws of supply and demand: "The year after the first discovery, the price of oil was $20 a barrel; at the end of the next year it was 10 cents a barrel, and sometimes a barrel of oil was literally cheaper than a barrel of water." Drake himself went broke soon after his triumph; he stayed in Titusville, serving as justice of the peace, until 1863, "after which he joined a New York brokerage firm that dealt in oil shares. By 1866 he had lost all his money and lived in poverty until 1873, when the state legislature granted him ... a small, lifetime pension for his service." Like a number of those who would follow him in the search for prosperity from petroleum, Drake received remarkably little wealth from his discovery, dying in obscurity in 1880.[9]

The Pithole Plummet

An even clearer rags-to-riches-to-rags story comes from a community a few miles to the southeast of Titusville, which today exists only as a footnote—the picturesque-sounding town of Pithole, Pennsylvania. Pithole, which appears to have been the world's first community to experience a complete energy boom and bust, no longer exists on most maps, but signs at the Pithole Visitor Center still tell the story of "Pithole, the fabulous boomtown":

> On the flats at the bottom of this hill there was a tiny farm in January, 1865. A well was being drilled. On January 7, the well began to flow oil, a lot of oil! Men rushed here to drill wells of their own. The farm was sold, divided into lots, and Pithole City appeared by summertime. By September, as many as 15,000 people lived on this hillside. Over 50 hotels had been built, and stores, banks, offices and saloons filled the land half-way down the hill. By the next January, some of the wells had played out, and Pithole began to die. Fires burned much of the town, but people leaving quietly to go to richer oil fields led to the rapid decline of Pithole. After a few years, the land at the foot of this hill once again was only farmland along Pithole Creek.

Across what used to be the streets of Pithole, things are mainly quiet today. If winds are calm, it is possible to hear the occasional backfiring of the "Drake engine" a few miles away, but at the site, the only other sounds usually come from the buzzing of the insects, the chirping of birds, and the murmurs of an occasional tourist. The quiet is a marked contrast to what must have been the hustle and bustle of Pithole's boom days, but the streets are still marked with respectful wooden posts. Near the posts marking the intersection of Prather and Fourth

Streets, a sign fills in visitors on "Pithole's oil production and land speculation":

> The Thomas Holmden farm future site of Pithole was a poor buckwheat farm in 1864 when the U.S. Petroleum Company leased a part of it to drill a well. When the well came in, the value of the Holmden farm jumped to $150,000 as it was sold to a pair of speculators. The farm was sold again for $1,300,000 and again for $2,000,000. Then the boom ended. The county paid $4.37 for the Holmden farm in 1878.

For a period of roughly 500 days, according to one account, Pithole was "the most legendary of oil-rush cities. At its peak, in 1865, it had 10,000 inhabitants.... It was the centre of oil communications: the Pithole post office was said to be the third busiest in the States."[10]

What is more likely is that Pithole's was the third-busiest post office in the single state of Pennsylvania, but even at that, there is little doubt that its level of activity was phenomenal. A forest of oil derricks sprouted up in an area of just a few hundred acres along Pithole Creek, and the underlying formation provided drillers with a brief bonanza of 3.5 million barrels of oil. Its many hotels provided their patrons with delicacies such as oysters brought in fresh from Baltimore, some 200 mountainous miles away. Although the town never got around to constructing a sewage system, it did have the world's first commercially successful crude oil pipeline—5.5 miles of 2-inch wrought iron pipe, which started carrying crude oil on October 10, 1865, at the rate of 81 barrels per hour, connecting the then-booming oil field to nearby communities such as Oil City and Oleopolis.[11]

In addition, Pithole left us with two sets of lessons, involving some of the most and least fortunate aspects of oil. On the more fortunate side, oil is one of the most valuable sources of fuel that humans have ever discovered. First and foremost, it is an extremely high-quality fuel, permitting far more work per unit of weight than almost any alternative fuels, such as coal or wood, or for that matter even whale oil, to say nothing of its value as a petrochemical feedstock. Second, there are additional advantages in the fact that oil and natural gas are fluids. Coal can be turned into coke and wood can be converted into charcoal, but even these relatively limited transformations lose many of the volatile compounds in the original substances that might otherwise have been available to provide energy. By contrast, liquids or gases are far easier to transform through chemical processes (such as refining and "cracking") to produce a variety of finished products. In addition, as Pithole's pipeline illustrated, fluids are remarkably easy to move around.

To this very day, we calculate quantities of oil in terms of "barrels"—42 gallons to the barrel—largely because it was in wooden barrels, and more specifically whiskey barrels, that Drake first shipped his oil. As the oil shippers of Pithole were the first to discover, however, the liquid form of crude oil allowed it to be shipped much more economically than could so many barrels of potatoes. Before the pipeline was completed, teamsters were hauling the barrels to the nearest railroad with horse-drawn wagons, at the cost of $3.00/barrel. Once the pipeline was completed, it could carry the same commodity at one-third the cost. In the words of another historic marker at Pithole, "This left thousands of rough, tough teamsters with horses to feed and no cargoes to haul." In response—and in a pattern that would be familiar to anyone who has seen television footage of pipeline sabotage in Iraq—oil companies found that they needed to invest heavily in security forces to protect the long, exposed stretches of their pipelines. Even so, the financial advantages of pipelines evidently still outweighed the disadvantages by a good margin. The world's first pipeline was soon followed by another and another, establishing a pattern of portability that continues up to the present.

The advantages of oil, moreover, continue past the ends of the pipelines. Even inside the final user's automobile, relatively simple pieces of equipment, such as a pump and a fuel injection system, can take the place of the hard-working man who was needed to shovel all of that coal into the firebox of an old-fashioned steam locomotive.

Today, oil is pumped from the ground into tankers for transportation to many areas of the globe. On arrival, oil is pumped from tankers into pipelines, which carry it to refineries. From refineries the downstream products are pumped into pipelines, barges, or trucks for distribution to wholesale and retail outlets, where consumers increasingly pump it themselves into their own automobiles. All of this occurs without the "handling" (loading, unloading by human labor), packaging, and displaying associated with most products we buy.

Imagine an agricultural industry that harvested products in the field mechanically, pulverized and mixed them in central locations, and pumped the resulting product to stations where consumers would fill their own containers; bypassing the entire human harvest, initial packing, wholesale transportation, cooking/mixing, repackaging, displaying in grocery stores, etc., and you can begin to get an idea of the advantage that oil offers as a product.

Furthermore, although oil and gas differ within limits in their chemical makeup, these differences can be accommodated by the refining process. So to use our agricultural analogy, suppose the process was limited to one crop to further simplify its production, processing, and distribution. It is precisely these properties of petroleum that have at least contributed to the enormous concentration of capital surrounding the harvest, processing, and distribution of petroleum, and the vertical integration (control from the well to the automobile) of the multinational corporations that are the result of that capital accumulation.[12]

Still, while oil has tremendous advantages, its disadvantages are notable as well. In particular, oil is after all a "fossil" fuel. It is ancient in its origins, and finite in its quantity. When it's gone, it's gone—and so are the communities and economies built on its extraction. In the technical literature, the pattern is sometimes called *overadaptation*. Pithole appears to have been the first community to experience that pattern, but it would prove to be far from the last.[13]

The reference to adaptation reflects something that ecologists have often pointed out—organisms always adapt to their environments, but adaptations that offer short-term advantages may prove to have more negative consequences in the longer run. There may even be a lesson in the fact that organisms in nature rarely seem to have evolved to a point of "maximum efficiency"—suggesting that there is a potential for an organism to be *too* finely tuned to a given ecological niche or given set of conditions. If not, the earth might still be the realm of the dinosaurs.[14]

Whatever their other characteristics, Pithole and its fellow boom-and-bust towns have been very efficient at—highly adapted to—the rapid extraction of energy resources. At least up until the time of the blowout, BP's managers probably thought they were being "efficient," as well—cutting "needless costs and delays," and significantly improving BP's profitability, rather than simply cutting corners. As Paul Ehrlich has often pointed out, though, we often can't know in advance just which costs are truly "needless." Removing one or two rivets out of an airplane, in Ehrlich's analogy, probably won't cause the plane's wings to fall off. Neither might the removal of the next rivet, or the next. The problem is that none of us know exactly how many rivets we can afford to remove—which is precisely why engineers have learned to "overdesign" their systems, or to make them significantly stronger than they "need" to be.

Over the years, however, the evolution of our society as a whole has left little such margin for error. Instead, we have in some ways become just as highly adapted to the available supply of fossil fuels as have the citizens of Pithole—meaning that we face the potential to experience something like the same fate if or when our finite fuel supplies were suddenly to run out.

In short, the dinosaurs may have left us not just a remarkable supply of energy, but also an important lesson. The more precise the adaptation to a given set of environmental conditions and resources, the greater may be the susceptibility to perturbations or unexpected changes. Even the BP blowout might not have occurred had it not been for the natural gas pressure problems, although the available evidence suggests that it would be stretching the meaning of the term to call those problems "unexpected." More broadly, the underlying lesson is one that may well apply to human communities and even societies: It may well be wise not just to avoid having too many eggs in one basket, but also to avoid having too many of our bets in one kind of oil barrel.

5 Barons and Barrels

For the barons of industry who took over American oil production in the era after the Civil War, the usual concern had little to do with overadaptation; they wanted to encourage the growth of oil consumption. The most influential oil baron of all—John D. Rockefeller, Sr.—was the man who played that game better than anyone else.

Rockefeller had not just a great deal of business acumen but, in many ways, a strong moral code. He worked very long hours, right from the days when he began his business career as a mere accountant in Cleveland, up to the point where he became the world's richest man, as well as becoming the first American to have more than a billion dollars in net worth. After accumulating all that wealth, he used much of it to support philanthropic causes, doing so with major and generally positive effects on medicine, education, and scientific research. His moral code, however, apparently never included any great concern about secrecy, ruthlessness, or collusion, whatever the laws of the day might have said.

Perhaps the most important key to Rockefeller's success was his decision to focus on refining capacity and transportation—plus his efforts to monopolize the former and manipulate the latter.

Following the end of the Civil War in 1865, and the start and end of Pithole a few years later, transportation in the United States meant railroads, which were beginning to emerge as the growing nation's most important industrial powers. Congress, state legislatures, and innumerable communities handed over hundreds of millions of acres of land to the railroads in the interest of securing access to the speed and other developmental advantages that railroads could offer. The nation's expanding railroad network also brought enormous personal fortunes to its owners—the Vanderbilt family, John Jacob Astor, Jay Gould, Edward Henry Harriman, and many of the other most glittering names of the Gilded age. Rockefeller, however, made his fortune by outsmarting the railroads, rather than owning them, and by getting better deals from the railroads than anyone else could get.[1]

The railroads are generally seen as having changed the fortunes of entire regions, making it possible for new centers of specialization and transition to emerge. During the era of the steamboats, for example, most observers expected St. Louis to be the gateway to the western states, given its location at the mouth of the same Missouri River that Lewis and Clark had followed on their way to the Pacific. As a result of the growing influence of the railroads, however, Chicago emerged as the de facto capital for the development of "the great west." Still, although Chicago was the key bottleneck for railroad transportation to and from the west, and although Pittsburgh was the site of the first oil refinery, the city that emerged as the early center of refining in the United States was a city not often remembered for its role in the oil industry today—Cleveland. The reason had to do with a simple difference that Rockefeller was quick to recognize: Pittsburgh was connected to New York City and other nearby markets by just one railroad, while Cleveland was connected by two.[2]

The Rockefeller oil empire thus began with a refinery in Cleveland, about 100 miles to the west of Oil Creek, which had a capacity of 600 barrels per day, or about 4 percent of the refining capacity of the entire United States at the time. Soon, the newly formed Standard Oil Company was negotiating with both of the railroads that served Cleveland, obtaining rebates that allowed Rockefeller's company to ship oil more cheaply than his competitors, providing higher profits, which could be used in turn to buy out any competitors that survived. In short order, Rockefeller swallowed up 22 of his Cleveland competitors, meaning that, within just a few years, he had managed to take over control of half of the world's refinery capacity.

By 1870, petroleum had been discovered and produced in a half-dozen states, including California—and Rockefeller and five other investors had set up the Standard Oil Company. Using the same strategies that had proved successful in Cleveland, Standard quickly moved outside of Ohio, and within less than a decade, Standard and its associated firms controlled 90 percent of U.S. oil refining capacity. In 1882, the company set up the Standard Oil Trust—an interlocking network of over 30 companies that became the first national monopoly in the United States. By the late 1880s, the Trust was operating in every one of the United States and was moving aggressively into the international market. Despite any number of attacks by reformers, the Trust and its successor organizations have played important roles in U.S. petroleum politics ever since.[3]

When it came to oil companies that were too big to fight or to swallow, on the other hand, Rockefeller quickly saw the value of cooperation, setting up another pattern that has continued to the present. That was rarely his first reaction; after encountering competition from a marketing firm that had been formed by two of Alfred Nobel's brothers and financed by the Rothschild banking interests of Paris, for example, Rockefeller at first created his own international firms, starting with the Anglo-American Petroleum Company in 1888. Before long, however, the giant oil interests controlled by Rockefeller, the Nobels, and the Rothschilds decided that they could all agree on one point—the reality of competition was far less attractive than the theory. Instead, they realized that, if no one of them could establish a literal monopoly, cartels might provide many of the same advantages, but with significantly less hassle. These three giants divided up the world market, which they would continue to control, with varying degrees of success, for over eight decades, until even these massive companies would bow to the newer kind of cartel that would be formed by oil-producing nations during the latter half of the twentieth century.

Dawning of a New Era

In some ways, things started to change for the oil industry around the time when the nineteenth century came to an end. The early warning sign may have come when Congress enacted the Interstate Commerce Law in 1887: Although that law focused on the railroads, it not only established the Interstate Commerce Commission but also established the policy that unfettered dominance by the robber barons might not go on forever. A few years later, due in part to the animosities toward the Standard Oil Trust that had been stirred up by investigative writers known as "muckrakers," the public outcry against the practices of Standard and some of the other major trusts led to the passage of the Sherman Antitrust Act, which was signed by President Harrison in 1890. Still, these laws were vague and were only listlessly enforced by then president Cleveland, and much the same was true under his successor, McKinley—who, in a pattern that might sound familiar by now, happened to have received $250,000 in campaign contributions from

Standard Oil. State-level regulation was somewhat more vigorous, with the Standard Oil Trust officially being dissolved by the Ohio courts in 1892, but that proved to be only a temporary annoyance; Rockefeller was able to regroup by consolidating his holdings under Standard Oil of New Jersey.[4]

Matters began to change more significantly in 1901, for two reasons. One is that, at the federal level, 1901 was the year President McKinley was assassinated and Teddy Roosevelt entered the White House. Unlike his predecessor, Roosevelt directed his Justice Department to prosecute both Standard Oil and the railroads under the Sherman Act. Even so, both Roosevelt and his successor, William H. Taft, were out of office by the time the legal battle had been fought all the way through the Supreme Court. It would not be until 1911—the same year when gasoline sales first topped those of kerosene, and three years after the first production Model T Ford was assembled—that the Supreme Court would order the Standard Oil Trust to be broken up into over a dozen companies. Two of the important companies, incidentally, were the predecessors of the companies that later became Exxon and Mobil—which would ultimately be recombined into the firm known today as ExxonMobil.[5]

But something else important also happened in 1901—and if the federal government's ability to change the behaviors of big oil firms was less than overpowering, the power of geology was another matter. That was particularly true of a patch of geology located near Beaumont, Texas, and named for the hill above it—some say because of the spindly pines that grew there, and others because the drilling rigs on top looked spindly themselves—Spindletop. That was the spot where, just a few days into the new century, on January 10, 1901, an independent driller named Anthony Lucas struck oil.

He was called an "independent" because he was independent from Standard Oil, and the state in which he was drilling was particularly unfriendly toward the Rockefeller empire, at least at the time. One of the reasons involved a character named James Stephen Hogg—a man who had served as the Texas attorney general and then governor during the 1880s and 1890s. A strong progressive, Hogg was self-proclaimed opponent of big trusts, which basically meant the railroads and Standard Oil. Under Hogg's leadership, the state developed a strong antitrust law, which effectively made it impossible for Standard Oil to operate legally in the state, as well as setting up the Texas Railroad Commission, which had been given such broad powers that it could not only regulate railroads but also later take over the regulation of the state's oil.

Tellingly, during the time when Hogg was developing his tools, few people realized that Texas had much oil. John Archbold—one of the directors of Standard Oil, and the man who became the company's president when Rockefeller stepped down—is even said to have laughed at the idea, offering to drink all the oil found west of the Mississippi.[6]

It would have involved a great deal of drinking. Spindletop proved to be the largest petroleum reservoir that had ever been discovered, and it came in with the kind of a bang that transformed the U.S. petroleum industry.

The bang was literal as well as figurative. Spindletop was the world's first big "gusher"— a well where oil came shooting out of the ground, driven by the same kind of pressure from natural gas that would later be seen in the disaster of the *Deepwater Horizon*. In a matter of seconds, the pressure blew about six tons' worth of four-inch drilling pipe out of the hole—followed a few moments later by a stream of oil some six inches thick, shooting more than 100 feet above the oil-drilling derrick, in a fountain of fortune that reportedly could be seen from ten miles away. The transformation of the industry, though, came from the even greater volume of oil that lay underneath and in similar underground formations nearby, called salt domes, which can be found all along the coastline of the Gulf of Mexico, and on both sides of the water line.

It took the crews nine days to shut off the gusher, by which time some 800,000 barrels had already come out—a "production" rate some 3,500 times higher than at Drake's first well in Pennsylvania. New investors came pouring into town before the gusher could be brought under control, and the population would triple in the next three months. In other ways, too, the scene bore more than a little resemblance to Pithole. More than 600 companies would be chartered there in the next year alone, and more than 285 wells would soon be actively working away on Spindletop hill. The areas around Spindletop saw the construction of storage facilities, pipelines, and more. In the words of *The Handbook of Texas Online*:

> Wild speculation drove land prices around Spindletop to incredible heights. One man who had been trying to sell his tract there for $150 for three years sold his land for $20,000; the buyer promptly sold to another investor within fifteen minutes for $50,000. One well, representing an initial investment of under $10,000, was sold for $1,250,000. Beaumont's population rose from 10,000 to 50,000. Legal entanglements and multimillion-dollar deals became almost commonplace. An estimated $235 million had been invested in oil that year in Texas; while some had made fortunes, others lost everything.[7]

In another pattern that would have important echoes later, the drilling at Spindletop showed far more concern about speed and profits than about industrial hygiene. Most of the derricks were located on plots that were just large enough to squeeze in a single platform, with planks being set up between derricks to allow people to get around. Soon, clean water would literally be more than a thousand times more valuable than oil, with water costing five cents a drink, or a dollar a gallon, and oil fetching three cents for each 42-gallon barrel—a cost difference of roughly 1,400 to 1.

The strong antitrust laws that James Stephen Hogg had put in place in Texas by that time didn't actually keep Standard Oil out of the state, but they did force the giant company to operate in secrecy. Given the quantities of oil involved, moreover, new players moved into the action, including Pittsburgh's Andrew Mellon and former governor Hogg himself. Perhaps because their entry into the industry was started by the discovery of a vast petroleum reserve, these new players did not share Rockefeller's reluctance to become directly involved in the production phases of oil. Instead, to compete with Standard's monopoly over refined or "downstream" products—oil, gasoline, and other finished products—the corporations that emerged in Texas, including the Texas Company (later Texaco) and Gulf Oil, did so as "vertically integrated" entities, meaning that they controlled their products at every stage of the process, from the exploration to the extraction to the refining and finally the retailing.[8]

Spindletop was monstrously larger than any oil field ever before discovered—large enough that it tripled the total of all U.S. production before that time. In combination with subsequent discoveries along the Gulf of Mexico, the new finds were enough to move the petroleum industry's center of gravity of the out of the eastern United States. The new discoveries also changed the culture of the industry, which soon came to be personified not by shrewd former accountants, but by legendary Texas "wildcatters," who got their name from bravely drilling in areas not previously known to contain oil. Some of the wildcatters, such as H. L. Hunt and Howard Hughes, became fabulously wealthy after finding oil at Spindletop, but the majority would die as poor as Edwin Drake. The. rough-and-tumble cowboy image that was actively promoted by some of the emerging leaders, such as Joseph Cullinan of the Texas Company, also became part of the emerging industry culture, where it would persist to the present.[9]

The oil reserves of Spindletop, by contrast, would not be so long-lasting. The field was massive, but it was still finite, so even though the wildcatters seem to have approached the extraction of oil as though there were no tomorrow—or perhaps precisely because of that fact—production soon began to decline. To quote again from *The Handbook of Texas Online*, "The overabundance of wells at Spindletop led to a rapid decline in production. After yielding 17,500,000 barrels of oil in 1902, the Spindletop wells were down to 10,000 barrels a day in February 1904." The coming of the end of Spindletop was not quite as instantaneous as the end of the Pithole field, partly because later and deeper exploration along the edges of the hill during the 1920s led to the discovery of a new, smaller gusher, which flowed at a controlled 5,000 barrels a day, leading to a second boom. Even so, however, most of the companies in the region would leave by the 1930s, searching for more profitable pastures elsewhere. By the 1950s, the field would be mined for sulfur, rather than for any further oil deposits. According to the Web site for the Spindletop–Gladys City Boomtown Museum,

> Today, the field looks like a wasteland. The sulphur mining left behind great sunken spots where the underlying sulphur was extracted. The land is barren, dotted with bunches of high weeds that hide the tops of rusted pipes marking spots where the wells used to be.... the Spindletop field is now classified a "stripper operation" by the Texas Railroad Commission.[10]

The Texas firms that started coming to prominence after the Spindletop discovery were more worried than Standard Oil had been about the fact that they were "producing" a finite commodity. That, however, did not convince them that they ought to remove the finite fuel more slowly. Instead, they did what almost any other oil company would do in later years—ask the federal government for financial backing or, more specifically, a tax break. In the case of the Texas firms, they argued that they should be allowed to take credit for the finite nature of their reserves when they were computing their income taxes.

On the surface, at least, they would seem to have had bad timing. It was after all in 1911—more than 40 years after J. D. Rockefeller started putting together the Standard Oil Trust, and 18 years after the Sherman Antitrust Act made his empire illegal—that the U.S. Supreme Court finally found the Trust to be in violation of the act. Still, in 1913, on the eve of World War I, Congress agreed to the producers' request for what became called a "depletion allowance," enabling the mining and oil interests to deduct 5 percent of the gross value of production from their taxes. Just a few years later—paying little evident attention to the 1911 Supreme Court decision—the federal government looked more kindly on a spirit of "cooperation" between oil companies and the federal government, as well as among companies that were officially competing with one another. This time, though, the impetus was not just the influence of a rich and powerful industry, but war. As Solberg put it, "Overnight, the industry, with President Wilson's support, was doing what the Sherman Antitrust Act forbade. Six years after the dissolution of Standard Oil's trust, its chief executive was in Washington helping direct industry's cooperation with government."[11]

As of 1900, 90 percent of U.S. energy consumption involved coal, and coal remained the nation's primary fuel until the 1930s. The Allied victory in World War I, however, depended heavily on petroleum, which fueled military trucks, planes, and ships. Perhaps partly for this reason, even though most industries (including coal mining) were subjected to heavy regulation during World War I, the oil industry was not. Instead, the Oil Division of the U.S. Fuel Administration established a "committee system" to encourage "voluntary increases in production and efficiency." The approach seems to have worked; after the end of the war, Britain's Foreign Secretary, Lord Curzon, said that the Allies had been "carried to victory on a flood of oil."[12]

Among its other effects, this cooperation led to a provision in the War Revenues Act of 1918, allowing oil producers to base the depletion on a "reasonable allowance" of either the cost or the fair market value of their discoveries. For most companies, this proved to be far higher than the initial 5 percent level, averaging between 28 and 31 percent of the companies' gross incomes. In another pattern that would be repeated later, this enormous tax break did in fact encourage exploration and drilling during the last year of the war, as intended, but the tax break did not come to an end when the war did. Instead, arguments would rage about whether the oil depletion allowance ought to be set at a different *rate*—being changed to 27.5 percent of gross income in 1926, for example, to alleviate the necessity to adjudicate each claim—but for the next half of a century, until the 1970s, U.S. tax policy would still effectively allow the oil companies to avoid paying taxes on roughly a quarter of the income they derived from producing oil wells.[13]

To be fair, policymakers may have felt that they had good reason for not dismantling the wartime measures: They and those around them appear to have been worried about running short of oil. As would be the case on any number of occasions in subsequent years, though, they responded by offering financial inducements for burning through the remaining U.S. oil reserves even more quickly.

The first federal warning of impending shortages of petroleum appears to have come from the U.S. Geological Survey (USGS) in 1906, at a time when there were only about 100,000 cars and trucks in the entire United States. After the end of World War I, USGS issued another warning, this one reflecting a bit more urgency. By that time, there were nine million cars in the United States—although the nation was still much more dependent on railroads than on automobiles, with nine out of ten trips being taken by train, and with fewer than one American in ten even owning an auto. Still, those nine million automobiles were consuming over three billion gallons of fuel per year by that time, and the agency warned that the longer-term outlook for domestic oil supply was "precarious," with perhaps just 20 years of oil remaining.[14]

A prudent manager with concerns about running out of a precious commodity might well take steps to conserve the supplies that were left, but in the long history of U.S. dependence on petroleum, that kind of thinking has been in remarkably short supply. Instead, America's political leaders got busy, giving away the nation's oil even faster.

President William Howard Taft was not just Theodore Roosevelt's successor in the White House—he had also been Roosevelt's personal choice for the office, having been groomed for the presidency by serving as secretary of state, secretary of war, and supervisor of the start of construction on the Panama Canal in 1907. After being elected easily in 1909, however, Taft proved to be substantially less skillful than Roosevelt as a politician. He alienated so many of his initial supporters that, in his campaign for re-election in 1912, he came in third, winning only the states of Utah and Vermont, and setting an all-time record for defeat by an incumbent American president.

One of the reasons he alienated so many initial supporters may have been that he often took his principles seriously. That seemed to be particularly true on matters of peace and war: In 1910, as the U.S. Navy was converting from coal- to oil-burning ships, President Taft set aside large areas of California and Wyoming as Naval Petroleum Reserves—areas that were believed to have significant oil-bearing potential and that were intended to provide the Navy with sources of additional fuel in case of future war-time emergencies.

Both Roosevelt and Taft were in office during what is often remembered today as the Progressive Era—a period of reforms that lasted from the 1890s to the 1920s, often with a focus on keeping corruption out of politics. The establishment of the Naval Petroleum Reserves was certainly consistent with the spirit of that time, and so too was a law that Congress passed ten years later, establishing the law that authorized but also governed the "leasing" of federally owned lands—the Mineral Policy Act of 1920. This new law changed some of the policies that had been established under the Mining Law of 1872, which had allowed the public lands themselves to be "patented," or in effect given away, to almost anyone who would stake a claim to the minerals rights and pay a small fee, generally amounting to $5 per acre or less. Under the new law, the federal government would offer a 10-year lease of the right to extract publicly owned deposits of petroleum and a range of other minerals, rather than turning over the deed for the land itself. In addition, the whole process was supposed to be carried out through a process of competitive bidding, with the leases going to the firm that would offer the government the highest bonus bid, or initial payment. In practice, however, the "responsible" federal officials often seem to have seen the requirements of the new law as something more like a minor annoyance.

The first president to be elected after the passage of that act, Warren Harding, took office in 1921. In his first year in office, he issued an executive order that transferred control of the Naval Reserves to the Department of Interior. The next year, his secretary of the interior, Albert Fall, chose to lease the former "Reserves" without worrying about the "competitive" part of the bidding. After years of investigations and lawsuits—during which key pieces of evidence seemed to keep disappearing—a federal court ultimately decided that Fall's decision had been influenced not just by America's need for more oil, but by Fall's apparent need for bribe money. Fall became the first Cabinet member in American history to be sent to prison for his actions in office, and one of the reserves—Teapot Dome in Wyoming, which got its name from a rock outcropping that vaguely resembled a teapot—gave its name to the biggest scandal in American history, at least at the time.[15]

Another supposed response to potential oil shortages that is far less well known, but no more virtuous, came with the effort to introduce lead into gasoline. Downplaying the health risks of lead that were already known at the time, savvy spokespersons from General Motors, DuPont, and, once again, Standard Oil, started to celebrate the discovery of the performance-enhancing properties of tetraethyl lead. In their public statements, at least, they claimed that this potent neurotoxin amounted to "a Gift of God," helping to improve gasoline mileage, at least modestly, during a time of evidently impending shortages. This was no easy sell: During just one week in late October, 1924, 80 percent of the workers at Standard Oil Company's experimental laboratories in Elizabeth, New Jersey, either died or were severely poisoned. Of the 49 workers at the plant, 5 died, and 35 others "experienced severe palsies, tremors, hallucinations, and other serious neurological symptoms of organic lead poisoning." There is no record about whether the companies made the argument that God sometimes works in mysterious ways.[16]

Progressive era or not, however, this was a time when the federal government had little ability to carry out scientific investigations of its own. Instead, that task was largely handed over to the oil industry—and the industry's studies somehow failed to find much evidence about the health damage that the lead inflicted after it was introduced into gasoline, or for that matter, into the air and soil of every major American city. As a result, generations of Americans learned to think of tetraethyl lead—or "ethyl"—as the premium grade of gasoline. Lead would spew out of America's exhaust pipes for the next fifty years, continuing until the introduction of the catalytic converter in the 1970s. Lead acts as a poison to catalytic converters, as it does to human brains, but it was the introduction of the catalytic converter that would finally provide the extra impetus for getting the lead out.[17]

Still, despite all of the help from Harding's White House and the addition of lead to gasoline, the more important factor in easing worries about the impending end of oil reserves had to do instead with an improved understanding of petroleum geology. The discovery of enormous new oil fields in Texas, Oklahoma, and California, in combination with growing production from oil fields in Mexico, the Soviet Union, and Venezuela, caused oil prices to drop significantly. By 1929—less than ten years after Congress authorized the leasing of federal lands, and in the same year when Harding's secretary of the interior, Albert Fall, was finally sent to prison—a newly elected President Hoover withdrew federal lands from leasing in an attempt to control overproduction. Later in that eventful year, the United States stock market crashed, securities lost $26 billion in value, and the world economy sank into the Great Depression. At that point, all worries about oil exhaustion seemed to evaporate.[18]

By 1931, with crude oil once again selling for just 10 cents a barrel, domestic oil producers took increasingly drastic steps to restrict production. The states of Texas and Oklahoma even passed state laws and stationed militia units at oil fields to prevent drillers from exceeding production quotas. Despite these measures, and despite federal efforts to intervene—with production restraints, import restrictions, and price regulations imposed by the National Recovery Administration—prices continued to fall. After the Supreme Court declared the National Recovery Administration unconstitutional, the federal government needed to be content with imposing a tariff on foreign oil.

With global production outstripping demand—and with serious international competition, including a price war in India, further depressing prices—the major oil firms turned once again to a "spirit of cooperation," at least with one another, if not exactly with the law. In September of 1927, the head of Shell Oil, Sir Henry Deterding, invited representatives of other major companies—including Walter Teagle, chair of Standard of New Jersey, plus John Chadman of Anglo-Persian, who may have been representing not just his company's interests, but also those of the British government—to a castle he had leased at Achnacarry, Scotland. Somehow, the idea of meeting in a castle must have seemed fitting. Safely ensconced behind the castle walls, these potentates of petroleum worked out the Pact of Achnacarry, which also came to be known as the "as is" agreement.[19]

At least under U.S. laws, this agreement was completely illegal, which may have been one of the reasons it was not announced with a major press conference. In effect, the pact created a world oil cartel, where the prevailing division of the world market was affirmed, a number of production quotas were set, and the need for competition was treated as a mere theory, rather than a legal requirement. Each market could be supplied from the nearest source, but a constant price was ensured. The cost, no matter where the oil came from or would be sent, was to be based on the current cost of the oil in the Gulf of Mexico market—effectively meaning Texas at the time—plus the cost to ship oil to that location from the Gulf. The agreement ensured orderly, carefully controlled marketing and development, and although it was uncovered by the Federal Trade Commission in 1952, the principles of the Pact remained largely in effect until the oil embargo of 1973–1974.[20]

6 Off the Edge in All Directions

Just a few years before the discovery at Spindletop, the world would experience another technological breakthrough in oil drilling, although it had a significance that only started to become more clear about fifty years later—the drilling of the first "offshore" oil wells, in 1898. The location was Summerland, California, less than ten miles away from where the Santa Barbara oil spill of 1969 would occur.

Summerland had already seen successful oil and gas wells onshore; a local story holds that, when baseball players wanted to continue their game after sunset, they pounded a pipe into the ground and simply lit the natural gas that came out of the top. Given the success of onshore wells, the next step was to continue building oil derricks along piers that jutted into the Pacific Ocean, making this the first location in the world where someone would drill through water to reach petroleum. As had been the case with the onshore oilfields at Pithole and Spindletop, though, the derricks were built just a few feet apart, and the field was soon depleted. Today, the breakthrough is commemorated by little more than a small bronze plaque.[1]

The offshore oil industry we know today was more strongly influenced by developments that took place in Louisiana and in Venezuela. If the Summerland oil derricks looked much like those of Spindletop, though, the developments taking place in Louisiana, in particular, would not; they would be more strongly affected by the nature of oil deposits that were found within salt domes. It was thus with the development of the Caddo Lake field in northwestern Louisiana, about a dozen years after the Summerland drilling, that the oil industry took its first step toward truly “offshore” drilling, unconnected to land.

By 1905, successful wells had been drilled close to that lake, and the Commissioners of the Caddo Levee District offered drilling leases under the lake itself in 1910. The top bidder was Gulf Oil, one of the companies spawned by Spindletop, which leased 8,000 acres. Developing those leases meant the need to deal with a pair of important technical problems that would later have implications elsewhere along the Gulf of Mexico.

The first involved the point just noted: In the middle of the lake, drilling and production had to be done without direct connection to land. The problem was solved, at least initially, through creatively adapting existing, land-based technology. Pilings were driven into the lake bottom, and platforms were constructed on top. Drilling equipment was hauled to the platforms by barge, and underwater pipelines were constructed to bring the oil ashore. This latter activity became a precursor to the extensive network of offshore and onshore pipelines that would eventually form an amazingly complex web across most of southern Louisiana.

The second problem was in some ways more important: the pressures in the underground reservoirs were higher than anything the oil industry had previously encountered. Many of the early wells blew out and burned uncontrollably. Even in this era, well before Earth Day, this unwelcome pattern led the state of Louisiana to pass laws regulating drilling procedures, as well as leading the General Land Office, in the U.S. Department of the Interior, to withdraw other public lands in the area from development. Engineers responded to the problems by developing early versions of a kind of equipment that would come to much wider public awareness in the *Deepwater Horizon* disaster—blowout preventers.[2]

The next major step forward in the development of offshore drilling technology took place in one of the largest oilfields ever discovered in the western hemisphere—Lake Maracaibo, in northern Venezuela, which ultimately produced approximately 4.6 billion barrels of oil. Drilling there needed to contend with a much larger and deeper body of water than Caddo Lake, since Lake Maracaibo was about two-thirds the size of Lake Erie and up to 120 feet deep—and it was connected to salt water. When the Lago Petroleum Company brought in the first well there, in 1924, the company soon discovered that the *teredo*, or shipworm, thrived in the brackish water of the lake—and it loved to eat the wooden pilings of the drilling rigs, which it could destroy in six to eight months. After experimenting with several varieties of local timber, to no avail, the company decided to use concrete pilings and to build a concrete base. Eventually, the concrete base was replaced with a steel deck, which eliminated the need for the central support pilings. As drilling moved into deeper water, the pilings were replaced with hollow, steel-reinforced concrete caissons—a technology that would later evolve into the kind of platforms that became the approach of choice in the North Sea and North Atlantic.[3]

The drilling under Lake Maracaibo also led to the further development of something that would become far more important back in Louisiana—movable technology, in the form of the steam-powered drilling barge. Rather than setting up and then later breaking down the drilling machinery on each platform, drillers in Lake Maracaibo mounted the power supply and the machinery on a barge, which could easily be moved from platform to platform, leading to considerable savings, as well as pointing the way to other new ideas.

By the 1920s, newly developing seismic techniques had located numerous salt domes along the Louisiana coast. Getting drilling equipment to those domes proved to be more difficult than just finding them, however, because much of the land in southern Louisiana is almost as much a part of the ocean as it is of the land. It owes its existence to special deliveries from what, after all, we call the "muddy" Mississippi River, over thousands of years. For millennia, the river has delivered topsoil from the regions we now know as Kansas, Illinois, South Dakota, and beyond—billions of tons of silt in all—building up an ever-broadening swath of wetlands, the largest contiguous coastal marsh in North America, across most of the southern boundary of the state.

The environment of this region has long been a fluid one, figuratively as well as literally, with constantly shifting rivers, bayous, and other water bodies. As one mouth of the Mississippi would choke with sediment, creating an impediment for the waters upstream, the river would carve other channels, then others and still others, in a "braided" pattern. Historically, most of the flow was split between numerous dividing and rejoining channels, with few of those channels ever carrying more than 20 percent of the total flow of the ancient river. The result is a vast, open region, known by geologists as an alluvial or deltaic plain, that stretches across almost 200 miles of southern Louisiana—from the Chandeleur Islands, off the coast of Mississippi, on the east, to Vermilion Bay on the west.[4]

The silt, combined with the region's abundant rainfall, has long supported a riotous profusion of plant life, meaning the fine silt deposits contain tons of decaying vegetation, adding another reason that neither the traditional land-based drilling techniques, nor those evolving in Caddo Lake or Lake

Maracaibo, would work very well. The decayed vegetation absorbs water, and the silt acts as a lubricant, resulting in what one observer has called "highly aqueous organic ooze," with a consistency only slightly stiffer than bearing grease. Drilling site preparation proved to be extremely expensive, even in shallow open water—the silt layer was far too thick to allow any pilings to reach "solid" ground, and drilling machinery produces significant vibrations, so even exploratory platforms might require up to 250 pilings, creating literal forms of "sunk" costs that were particularly unwelcome in the case of dry holes. The soggy soil also made it both challenging and highly expensive to move equipment in the ways in which it would ordinarily be moved on firmer ground, namely by road or rail.[5]

The solution to this challenge involved an extension of the kind of floating mobility seen in Lake Maracaibo—a mobile drilling barge. An employee of one of the companies that had its start at Spindletop, namely the Texas Company—the one that would be known to later generations as Texaco—submitted plans for a mobile drilling barge in 1932, only to discovered that a man named Louis Giliasso had already been issued a patent for a "submersible drilling barge." The idea was to take the barge to the drilling site, at least in shallow water, and sink it, providing a stable base for drilling. The elevated drilling derrick could be positioned over a slot in the bottom, and after the drilling was completed, the water could be pumped out of the barge, which could then be moved to a new location. It took until the next year to locate Giliasso, who by that time was operating a bar in Panama, but he and the Texas Company reached an agreement for the use of the patent.[6]

The *Giliasso* (as the first barge was appropriately named) was a success, starting a revolutionary expansion of drilling in this challenging environment. The barges would soon get larger, more powerful, and more ingenious, permitting ever-deeper drilling. One notable innovation involved breaking the operation into two barges, with half of the drill slot being built into each barge. Each barge could be made the size of the older submersible rigs, effectively doubling the available space once the two were connected together at the drilling site, but retaining the smaller size of the older rigs, hence fitting into existing canals.[7]

In a pattern that may require renewed attention today, however, the legal framework for controlling coastal drilling evolved a bit more slowly than did the technology. Those canals were dug by another important kind of mobile equipment—barge-mounted draglines—which were used to carve an ever-growing network of canals and pipeline corridors for moving the drilling barges into the marshes and for getting out the oil and gas. Unfortunately, in an ecological version of the proverbial "death by a thousand cuts," the barges also proved to be highly effective in destroying wetlands.

In the four decades between 1937 and 1977, approximately 6,300 exploratory wells and over 21,000 development wells would be drilled in the eight Louisiana coastal parishes (counties), mainly in wetlands or inland water bodies. Marshes were not seen as being terribly valuable real estate at the time, so drillers were allowed to dig an almost unlimited network of barge canals and pipeline corridors, resulting in the loss of over 190 square miles of land due to canal surface alone, along with an unknowable but substantial loss of adjacent lands. In many cases, the canals and pipeline corridors allowed salt water to reach and kill salt-sensitive plants in formerly freshwater marshes. In other cases, the dredging created continuous "spoil" piles along the banks of the canals, keeping the marshes from draining normally after heavy rainfalls, thus killing plants that could survive short periods of inundation but not long ones. As noted by one Louisiana resident we interviewed in the early 1990s, "It's not the hundred acres you take to build the canal that's so devastating, it's the 10,000 acres you destroy because you build the canal, and then you block the bayous and streams, and you block the tidal exchange." Overall, although there is no definitive figure on the overall level of damage created by coastal oil-exploration practices, that damage is clearly substantial.[8]

Some of the first efforts to develop a legal framework to control the leasing and its associated environmental impacts came from California, which had been leasing offshore tracts in the Santa Barbara channel, but which found by the late 1920s that applications for offshore tracts were getting out of hand, partly because the state lacked the legal foundation to control the industry. In 1929, the state legislature repealed the California State Mineral Leasing Act, effectively curtailing offshore leasing. Not until the State Lands Act was passed, in 1938, giving California legal control over offshore leasing, was limited offshore development resumed.[9]

Most of the drilling, however, was in and around the state of Louisiana, and in Louisiana, the main reasons for legal attention had to do with economics, not ecology. The wetlands are teeming with life, but one aspect of that fact is that they are filled with what a local resident once called "things that will stick you, sting you, stab you, and bite you." The well-known humor columnist Dave Barry was speaking of Florida, not Louisiana, in one of his columns, but he might well have been representing the views of many residents of southern Louisiana in earlier decades when he described the Everglades as "an enormous, wet, nature-intensive area that at one time was considered useless, but which is now recognized as a vital ecological resource, providing [the state] with an estimated 93 percent of its blood-sucking insects."[10]

Across most of the southern end of Louisiana, this extensive array of coastal marshes makes it almost impossible to get to within 15 to 20 miles of salt water without a helicopter or a boat. In an environmentally insensitive age, when the marshes seemed almost impossibly vast, few residents of the state were worried about the dredging of a few more canals, and then a few more. Most residents did not become concerned until the cumulative impact of the canals finally started to become too clear to ignore, roughly half of a century later.

When the state of Louisiana offered to sell the first lease for truly offshore drilling, in 1945, the only bidder was the Magnolia Petroleum Company—another Texas production company, but in this case, a company that had been bought in 1925 by Standard Oil of New York and that was later to become part of Mobil, then ExxonMobil. The following year, using a mixture of wood and steel pilings, Magnolia constructed a platform on their newly acquired, 149,000-acre tract, five miles from the nearest land. The site was south of Morgan City, which is still the home of a monument that commemorates the historic effort. The platform was designed to withstand 150-mile-per-hour hurricane winds and a deck load of 2.25 million pounds. In a move that would be repeated many times over in the years to come—and that would contribute significantly to positive relationships between oil and the fishing industry—Magnolia hired local shrimp boats to deliver supplies, equipment, and even the work crews, who were rotated back and forth from "floating hotels" that were anchored behind Eugene Island, ten miles away.[11]

That first well was an economic failure—a dry hole—but it was a technological success, demonstrating the feasibility of oil exploration in the open Gulf. It also helped to contribute to what became known as "The Tidelands Controversy," which was not the kind of controversy associated with offshore oil drilling today. It involved cash.

The states that were experimenting with offshore oil exploration at that time—Louisiana, Texas, Florida, and California—all worked on the assumption that they, the states, owned the sea bottoms. There were those in the federal government, however, who did not agree. Particularly notable was the Secretary of Interior under Franklin Roosevelt, Harold Ickes, who had encouraged the U.S. Senate to pass a resolution in 1937, directing the attorney general to assert federal ownership of offshore lands. The House of Representatives took no action on the matter, however, and World War II soon overshadowed the issue. After the end of the war, though, Ickes continued to push for federal ownership of offshore lands, and President Truman was sympathetic to federal ownership. In 1945, President Truman issued Executive Order 9633 (Federal Register 12304 (1945); 59 Stat. 885), asserting federal ownership of the offshore oil lands. In addition, Ickes persuaded Truman to initiate a suit against California in the U.S. Supreme Court.

The states argued that they owned mineral deposits in the waters adjacent to their coasts as remnants of their jurisdiction over the "marginal sea"— the three-mile strip of sea adjacent to a nation's coast—as original colonies. The federal government argued that the marginal sea concept did not arise until after the Revolutionary War. In spite of the Truman Proclamation and the impending suit, but in the absence of clear legal prohibitions, the state of Louisiana simply moved forward with its efforts to lease offshore lands, and the state of Texas soon followed.

The legal battles went on for years, but the outcome is one that will come as little surprise to anyone who is familiar with other such fights. The conflicting claims between the states and the federal government were ultimately settled by the U.S.

Supreme Court, in a series of decisions handed down between 1947 and 1950 that are generally remembered as the Tidelands Cases (see for example *United States v. California* 1947). The decisions established the legal rights of the federal government over all offshore lands.[12]

That, however, was not the end of the story. Given the strength of feeling in many coastal states—and in the oil companies—the Tidelands controversy became a major issue in the 1952 presidential campaign. The Democratic candidate, Adlai Stevenson, came down clearly on the side of federal ownership, while the Republican candidate and ultimate winner, Dwight Eisenhower, supported the states' claims. "Feelings ran highest in Texas, where the Tidelands issue topped all others. When Democratic candidate Adlai Stevenson came out for federal control in the face of crowds bearing placards 'Remember the Alamo and the Tidelands Oil Steal' [the state's Democratic] Governor Allan Shivers announced that he could not support the party's candidate.... The oil of the tidelands had much to do with detaching Texas from its historic allegiance to the Democratic party."[13]

The fact that oil companies launched a major lobbying effort may also have contributed to the outcome. In light of later federal lenience toward the oil companies, the expectation may be a bit difficult to believe, but the oil companies felt that the states would be even more lenient in their regulatory approaches than would the federal government. One industry-oriented publication tried to make the issue an even larger one, ignoring the fact that the undersea lands were already "nationalized," or owned by the public, and instead warning ominously that "a socialist-minded administration could easily nationalize all property and socialize the country."[14]

After the inauguration of President Eisenhower in 1953, and with his encouragement, Congress moved quickly to change the laws of the land, passing two pieces of landmark legislation that have continued to shape federal policy to this day. The first was the Submerged Lands Act of 1953, a compromise that assigned to states the title to offshore lands that were within three miles of the shoreline. Two exceptions involved Texas and the west coast of Florida, where the Supreme Court subsequently ruled that the states had held title to three marine leagues (approximately 10.4 miles), as sovereign nations, before they were admitted to the Union.

The second piece of legislation, which built on the first, focused on the so-called "Outer" Continental Shelf, or OCS. The reference to the "Shelf" reflects the fact that, along much of the continent—including the Gulf coast—there is an almost literal shelf that extends beyond the margins of the continent, underlying relatively shallow waters. The "Outer" part of the title refers to the sea-bottom lands that are "out" beyond the limits of state jurisdiction, meaning more than three miles (or three marine leagues) offshore. This second law, the Outer Continental Shelf Lands Act (OCSLA) of 1953 (43 U.S.C. 1331 et seq.), authorized the secretary of the interior to offer leases for oil and gas (and salt and sulfur) on the Outer Continental Shelf, through competitive bidding, and subsequently to administer the leases.

If these two laws marked important beginnings, however, a third development in that same year of 1953 involved an ending—the ending of the first century of oil, from 1854 to 1953. This third development is one that has received far less attention in most historical accounts, yet it may have been at least equally important. That year marked the closing of roughly a century during which over half of the oil in the world had come from a single nation. The nation that lost its petroleum preeminence in 1953, however, was not Iraq or any other nation in the Middle East. It was the United States of America.

The Other Gulf

In some ways, the beginning of the end of U.S. dominance in world oil markets can be traced to the same year that Spindletop seemed to be ensuring that dominance, namely 1901. That was the year when a British-born entrepreneur who had made a fortune in Australian mining operations, William Knox D'Arcy, negotiated a 60-year oil concession with the Shah Mozzafar al-Din Shah Qajar of Persia, obtaining exclusive rights to look for oil in a vast tract of territory that included most of present-day Iran. In the spring of 1908, D'Arcy managed to locate the first commercially significant oil deposit in the Middle East, and a year later, he founded the Anglo-Persian Oil Company. It was the start of the company that after its renaming in 1954, the world would come to know as British Petroleum.

Within a dozen years of his first discovery, oil exploration and development would become truly international, as well as much more massive. For the time being, however, it would not bring much power to the nations of the Middle East where the new deposits were starting to be found.

World production of crude oil had risen roughly one hundred thousand-fold by the start of the 1920s—from six thousand barrels a year in 1859, when Drake drilled that first well, to 689 million barrels in 1920, the latter figure being approximately the annual capacity of the Trans-Alaskan pipeline. One of the reasons that oil prices would have declined so significantly in the late 1920s, in fact, may have had something to do with the laws of supply and demand. Oil was discovered in what is now Iraq in 1927, in Bahrain in 1931, and in Saudi Arabia and Kuwait shortly thereafter.

With all of this new oil coming into the market, the supremacy of the Rockefeller empire had begun to be challenged not just by the newer Texas-based firms in the United States, but also by several international corporations. Chief of these was the Anglo-Dutch Shell Oil Company, which had moved into the west coast of the United States, as well as exploiting massive holdings and markets world wide. By the mid-1920s, in fact, Shell had become the world's largest producer. Another international competitor was the company that started the shift to the middle east, namely D'Arcy's Anglo-Persian Oil Company, which in 1934 would be renamed as the Anglo-Iranian Oil Company.[15]

In a further display of the new spirit of "cooperation" under the Achnacarry agreement, however—but in this case with the support of the U.S. government—five of the major U.S. oil companies entered into a consortium with British, Dutch, and French interests, forming yet another multinational firm, the Turkish Petroleum Company, which would later be renamed the Iraq Petroleum Company. In short order, both the original companies and the new entrants into the Middle East were entangled in a complex series of agreements and interlocking directorates that almost completely controlled the Middle East's oil production.[16]

Acting in a way that continued to emphasize "cooperation" over irritating legal requirements, the consortium members agreed not to act independently in an area covered by the "redline agreement" of 1928—so named because one of the negotiators literally walked to the map and drew a red line around the area the agreement would cover. In a detail that would prove significant in later decades, that area included Turkey, Cyprus, Lebanon, Syria, Jordan, Iraq, Saudi Arabia, and most of the remainder of the Arabian Peninsula, but it did not cover Kuwait or Iran. Although the red-lined areas were later found to contain the majority of the world's proven crude oil reserves—as well as a flash point for foreign policy disputes and later wars, connecting in other ways to the industrialized nations of the world—the development of those oil fields would have to wait until after the end of another war.

War and Peace, Power and Petroleum

For almost the entire decade of the 1930s, the major focus of energy policy attention was the widespread concern over over-supply, but when World War II broke out in Europe in 1939, policy attention again switched quickly to a desire for increased production. To put the matter simply, World War II was fought with oil, and over oil. Japanese strategy throughout hinged on increasing the imports for their island nation; some analysts have concluded that the Japanese bombing of Pearl Harbor may have been a response to the fact that U.S. exports to Japan were halted in 1941.[17]

Much the same was true for Germany's efforts—both during the war and during the period leading up to it. In charges that were vehemently denied by Standard Oil, a number of observers charged that Germany's preparations for war were carried out with significant help from Standard Oil. In the words of Sutton:

> In 1934 for instance about 85 percent of German finished petroleum products were imported. The solution adopted by Nazi Germany was to manufacture synthetic gasoline from its plentiful domestic coal supplies. It was the hydrogenation process of producing synthetic gasoline and iso-octane properties in gasoline that enabled Germany to go to war in 1940—and this hydrogenation process was developed and financed by the Standard Oil laboratories in the United States in partnership with I. G. Farben.[18]

At least during the war, though, U.S. oil companies were seen as unquestioned patriots. Even the secretary of the interior at the time, Harold Ickes—the same man who had fought for federal control of offshore oil reserves—proved to be quite friendly to the oil industry during the war itself. After he was named "Petroleum Coordinator for National Defense," Ickes arranged for the industry to be "secured immunity from a considerable amount of anti-trust regulation." He also pushed for funds from the Reconstruction Finance Corporation to pay for the construction of two major pipelines from mid-continent oil fields to the eastern seaboard, doing so partly in response to the increasing danger to oil tankers in the Gulf of Mexico from German submarines. The major focus of oil production, however, was for the war effort itself. Overall, the Allies used roughly nine billion barrels of oil during the war, and eight of those nine billion barrels came from the United States.[19]

At least in terms of petroleum, the U.S. energy supply would never again look so rosy.

The United States had been the country where the modern-day version of drilling for oil originated, in 1859. It was also the country that, for most of the century that followed, produced over half of the oil of the world. A few decades later, U.S. citizens would fret about the Organization of Petroleum-Exporting Countries, or OPEC, but up through 1953, the United States itself had been a virtual one-nation OPEC. In many ways, though, the signs of change should have been clear by the time that Americans and their Allies were holding their victory parades, celebrating the end of WW II.

Part of what both world wars had shown was that petroleum meant power, and by the end of the second one, U.S. petroleum supplies were shrinking rapidly. The other nations that were starting to be major suppliers of the world's oil, however, were not yet enjoying much power from their petroleum—and they were starting to be increasingly unhappy about that fact. In Iran, in particular, there was growing pressure for the revenues that could come from sitting atop a precious and rapidly dwindling resource. The pressure was resisted by Iran's pro-Western prime minister, Ali Razmara—but only until he was assassinated. After that, Mohammed Mossadeq was named as prime minister, and Iran's oil industry was nationalized.

Understandably, while this development may have been popular within Iran, it was much less popular among western politicians or petroleum companies. Initially, the Anglo-Iranian Oil Company withdrew its management team and organized an effective boycott of Iranian oil, but the intrigue did not stop there. Instead, in the same fateful year when the United States set up its offshore drilling program, 1953, the new U.S. president, Dwight D. Eisenhower, authorized the Central Intelligence Agency to carry out "Operation Ajax." With support from the British government, the shah, and the Iranian military, this CIA-organized coup forced Mossadeq out of office in August 1953. Mossadeq was replaced by a pro-Western general and the shah of Iran, who returned to the country, abolished its democratic constitution, and established the kind of rule that was much more to the liking of western oil companies.

It was during all this intrigue, in 1954, when the Anglo-Iranian Oil Company changed its name to British Petroleum.[20]

With the same year that marked the arrival of British Petroleum, the United States would permanently relinquish its role as the world's major oil supplier. It would not, however, relinquish its role as the world's major oil *consumer*. "By 1956 Americans made up 6 percent of the world's population yet owned two thirds of its cars." Ignoring geological reality, but not political reality, American leaders did all they could to encourage the further growth of oil consumption, making a key series of choices that consistently pushed Americans toward an increased dependence on the very kind of fuel that, by that time, the United States no longer controlled.[21]

The main vehicle, literally as well as figuratively, was the automobile. Auto sales had been slow during the Great Depression,

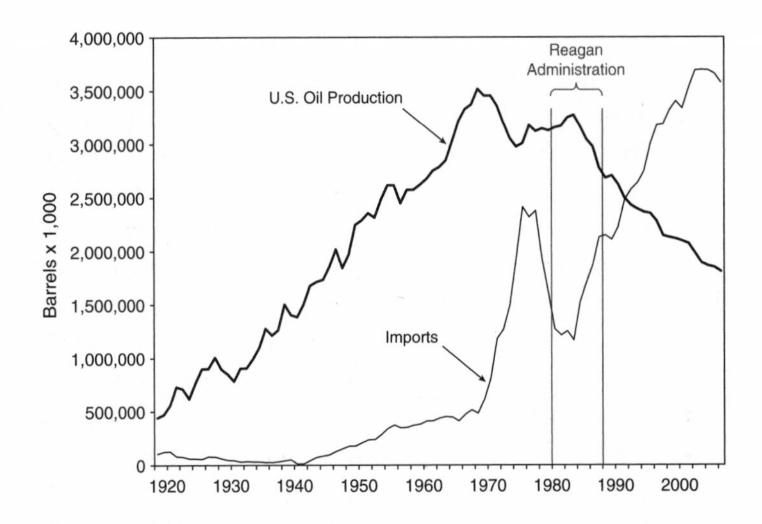

Figure 6.1
U.S. percentage of world oil production, twentieth century

and wartime production controls strictly limited vehicle sales through 1945. After that, however, sales exploded. In 1940, the nation had 27.5 million registered vehicles, in a population of 132 million residents, or a ratio of just over one vehicle for each five humans. By 1960, a population of 179 million residents would be driving 61.5 million vehicles—more than one vehicle for every three residents of the nation, or more tellingly, more than one vehicle for every household in the country. Yet the boom in automobile use, and in petroleum demand, was driven not just by an abstract American "love affair" with the automobile. It was also driven by the kinds of policy choices that would have inspired few objections from two of America's most powerful industries at the time—petroleum and automobile manufacturing.

In many ways, the policy choices of the 1950s were merely an extension of the patterns that had been established in previous decades. As early as 1934, for example, the Federal Housing Administration had been given the laudable goals of providing depression-era Americans with jobs and improving the nation's housing stock. The legislation that created the agency, the National Housing Act, gave no overt indication of any bias against central cities. The agency's leaders, however, expressed concerns that the Americans who were most in need of better housing, namely the poorest ones, might default on their house payments. As a result, the agency effectively required the building of houses that were of a size and quality desired by more affluent citizens, while offering unfavorable terms for, and hence actively discouraging, the building of multifamily units, or even the making of repairs. By 1939, the agency's *Underwriting Manual* "taught that 'crowded neighborhoods lessen desirability,' and 'older properties in a neighborhood have a tendency to accelerate the transition to lower class occupancy.'"[22]

Rather than seeking to serve all Americans, let alone those who were poorest and most ill-housed, the Federal Housing Administration "was extraordinarily concerned with 'inharmonious racial or nationality groups.'" The bias, moreover, showed up in ways that were tangible as well as attitudinal. A careful study of the St. Louis metropolitan region, for example, showed that from 1934 to 1960, the federal government invested more than five times as much loan money in developing the suburbs around St. Louis as in the city itself—$559 million vs. $94 million. On a person-by-person basis, the disparity was even larger, amounting to a factor of more than six times as much money for the suburbs as for the city. As the author of that study noted in an earlier article in a respected journal, *Records of the Columbia Historical Society*, "The FHA was helping to denude St. Louis of its middle-class residents." Similar patterns were played out from sea to shining sea.[23]

It was in the exuberant era after the end of World War II, however, when the most fateful policy choices were made. All across Europe and Japan, investments were made in modernizing railroads and other forms of mass transit, sometimes rebuilding from the rubble of wartime destruction. In the United States, by contrast, most major cities had working streetcar systems—often aging and in need of upgrades, but untouched by war damage—that were systematically torn out over the next 25 years.

Many factors were involved, but one of them involved the efforts of National City Lines—a company with "no visible connection to General Motors," except that "the director of operations came from a GM subsidiary, Yellow Coach, and members of the Board of Directors came from Greyhound, which was founded and controlled by General Motors." By 1946, National City Lines controlled public-transit systems in over 80 cities across the United States. The trolley lines in those cities were torn out, being replaced with "modern" buses that happened to be manufactured by General Motors. The Justice Department indicted nine corporations and seven individuals on antitrust charges, and on March 13, 1949, all were convicted on one count of conspiring to monopolize a part of the trade and commerce of the United States. None of the guilty parties did any jail time, however, and the fines were negligible.[24]

Yet the tearing out of mass-transit systems was scarcely the only set of fateful policy choices to have been made around the same time. In 1946, for example, only 17 percent of American cities required businesses or apartments to provide parking. Five years later, 71 percent did. By the early years of the twenty-first century, the American Planning Association's compendium of regulations, "Parking Standards," would be more than 180 pages long, outlining "minimum parking requirements" for nearly every economic activity imaginable. The process, moreover, helped to feed a self-reinforcing spiral of hidden costs and increasing sprawl. In 1961, when Oakland, California, started requiring apartment buildings to have one parking space per apartment, housing costs per apartment increased by 18 percent, and urban density declined by 30 percent. Today, shopping malls try to offer enough "free" spaces for the busiest shopping day of the year, guaranteeing that many acres of impervious parking surfaces will simply sit there—contributing to sprawl, the runoff of pollution, and urban "heat island" problems—for the other 364 days of the year. The patterns have been repeated again and again, quietly contributing to sprawl problems in nearly every region of the country.[25]

One of the best-known of the nation's policy choices, however, involved the building of the interstate highway system—a process that, ironically, also started at roughly the same time as the United States lost its distinction as the source of half of the petroleum in the world. The Federal-Aid Highway Act of 1952 authorized the first funding specifically for interstate highway construction, although the amounts of funding involved were tiny. It was only after the election of President Eisenhower, and under his active promotion, that legislation and funding moved forward.

The lobbyists pushing for the construction of the interstate system "ranged from the Automobile Association of America to rubber manufacturers," but aside from the president himself, perhaps the key person was retired general Lucius Clay, "who also happened to sit on the board of General Motors." Eisenhower appointed Clay as the head of a commission to evaluate the nation's highway needs, and in early 1955, the commission reported that an interstate system was "an urgent necessity" as a way for the nation "to evacuate its cities in case of a Soviet missile attack."[26]

The "national defense" argument seemed to close the deal. By the next year, Congress passed the Federal-Aid Highway Act of 1956, popularly known as the National Interstate and Defense Highways Act, which "served as a catalyst for the System's development and, ultimately, its completion.... It also called for nationwide standards for design of the System, authorized an accelerated program, established a new method for apportioning funds among the States, [and] changed the name to the National System of Interstate and Defense Highways." In a point that was surely not lost on officials from state and local governments, finally, the bill effectively repealed the long-time expectation that local and state taxes had to pay for half of the cost of building highways, "set[ting] the Federal Government's share of project cost at 90 percent."[27]

The decision may also have been influenced by the considerations that were not quite so distant as the threat of a Soviet missile attack. At the time, 1.5 million workers were employed in vehicle manufacturing, petroleum production, and the building of roads.

In the exuberance of postwar American power and prosperity, it may have been difficult to foresee the significance of policy choices that pushed the United States toward an ever-greater dependence on petroleum. Those choices, however, proved to be particularly fateful, given that they were made at the very time when U.S. petroleum dominance had already ended. Fifty years later, General Motors would be teetering on the brink of bankruptcy, and the United States would be coming perilously closer to the end of its remaining petroleum supplies. By then, however, it would be difficult to reverse the choices that were made in the first few years after the end of World War II. The building of massive urban parking lots, the completion of the interstate highway system, and the loss of former mass transit systems, would remain as enduring facts of hardware, even though the next 50 years would look quite different from the 50 that had preceded them.

All in all, in fact, at just about the time when Congress was passing the National Interstate and Defense Highways Act, the reference to "defense" also came to take on a more ironic meaning. If the predominant pattern in the wars during the first half of the century had been for U.S. oil resources to provide the margin of victory, the pattern during the last few decades of the twentieth century came increasingly to look more like the spending of Defense Department dollars to seek or preserve U.S. access to oil supplies from other nations. Elsewhere in the federal government, though, efforts were still underway to speed up the removal of the remaining oil reserves of the United States.

7 “Energy Independence”

More than two decades ago, during the last period when large numbers of Americans were paying much attention to their energy uses, a book by Gibbons and Chandler noted, “While it may not be necessary to go all the way back to creation to begin an analysis of energy, more is called for than the usual ‘Beginning with the Arab Oil Embargo of 1973.’” Still, aside from those who worked for the oil industry, most Americans seem to have paid little attention to petroleum until what is often remembered as “the” energy crisis—the sudden spike in energy prices in response to the oil embargo of 1973–1974. Unfortunately, much the same is true of those who analyze energy policy for a living: The usual tendency is to see the first real statement of U.S. energy policy as being the one that then president Richard Nixon issued in response to the embargo—“Project Independence,” so named because it was supposed to bring independence from imported oil by 1980.[1]

Up until 1973, the world’s most important oil embargo, economically speaking, had been the 1959 *U.S.* embargo on *foreign* oil, which had significantly reduced the revenues flowing to other oil-producing nations. Further reductions had come after the oil companies engineered price cuts in crude oil in August of 1960, and still more followed from the increasing availability of Soviet oil around that same time. It was in the effort to gain control of the rate of production and price of their oil that General Oassem of Iraq invited representatives from Kuwait, Saudi Arabia, Iran, Qutar, and Venezuela to a meeting, in September of 1960, where the Organization of Petroleum-Exporting Countries was born.[2]

In the decade following its inception, OPEC had little effect on the production or price of oil, but that was changed by the rise of Colonel Qaddafi to power in Libya in 1969. By 1973, Qaddafi had nationalized the holdings of major multinationals, and unlike what had happened when the Anglo-Iranian Oil Company had been able to punish Iran for nationalizing its holdings in the 1950s, Qaddafi's efforts made it clear that, this time, that the nations that had the oil might actually be in a more powerful position than the nations and corporations that wanted it. Qaddafi succeeded in forcing the oil companies to negotiate a new price for Libyan oil, providing a new and more activist model for OPEC.

The power of that new model would be demonstrated during the last few months of 1973, when, in retaliation for their support of Israel during the 1973 round of the Arab-Israeli war, the United States and the Netherlands would be the target of an embargo from the Arab nations of OPEC. The embargo actually lasted just five months—from October 18, 1973, to March 18, 1974—but it provided the breaking point for the multinationals' control of the world oil market. In the three-month period from October 1973 to January 1974, the price of Arabian Light crude oil increased from $5.12 to $11.65 per barrel; this increase, moreover, came on top of the doubling of price that had already taken place, starting from the price of $2.59 per barrel in January of 1973. As American motorists were quick to notice, these changes meant that, after decades of basically stable or even declining prices, the cost of gasoline basically quadrupled in the space of a single year. As energy analysts were a bit slower to notice, there was also an important implication to be seen in the fact that these increases were dictated by the oil-producing nations, rather than the oil companies. As one observer would put it, roughly a dozen years later, the OPEC embargo went a long way toward turning the multinationals into "suppliers of technology and marketing agents for OPEC oil."[3]

That was not what the beleaguered President Nixon wanted to see. Instead, his "Project Independence" was intended to provide a quick fix to the shortages and price increases—or perhaps a return to the exuberance of the early 1950s—by inspiring a dramatic increase in U.S. oil production. His plans, though, ran into the usual problem—oil is not "produced" by today's humans, only extracted from finite reserves. By 1973, the United States was already decades past the time when it would have been possible to take care of American petroleum demand with American petroleum supplies. Instead, almost all of the major domestic petroleum deposits that were reasonably accessible had already been tapped, and they were well on their way to being drained. There were really just two more places to look—"frontier" regions for offshore oil, meaning those that had not yet had much exploration and development, and the state that sometimes calls itself the last frontier, Alaska.

If one were to ask what coastal region of the United States differs the most from the steamy wetlands along the Gulf coast, it would be hard to come up with a sharper contrast than north slope of Alaska. Receiving so little precipitation that it can technically be classified as desert, this region sits mainly on permafrost—land that, starting just a few inches below the surface, is frozen solid, year-round. It is located well above the Arctic Circle, meaning that there are times of the year when the sun never sets, and other times when it never rises above the horizon. As different as they may seem, however, these two edges of the continent—one to the south and east, the other to the north and west—are united by the fact that, deep below the surface, both have major, world-class deposits of oil.

Oil seeps in the state we know today as Alaska were first discovered by Russians, soon after the beginning of the nineteenth century, but the seeps were far from world markets, and local uses were limited mainly to the making of bricks. The first claims for oil drilling were made around the time of the Klondike gold rush, in 1892, but filing claims is not the same thing as finding oil, and all six of the early wells were unsuccessful. Alaska's first oil-producing well was drilled in the same year the gusher was discovered at Spindletop, 1901, but the North Slope of Alaska received little attention at the time. Although Charles Brower had reported North Slope oil seeps to the U.S. government around 1890, his report was forgotten until seepages were discovered again in 1917, and the first claim in the far north would not be staked until 1921. Even then, because of the tremendous logistical problems associated with drilling in such a cold and remote environment, no real oil development followed. Instead, in 1923—the year after Albert Fall had been bribed to lease the oil reserves in Colorado and California—Naval Petroleum Reserve No. 4 was established on Alaska's north slope. The reserve consisted of 23 million acres around Point Barrow, where seepage had been observed. It was not until 1944, with the prospect of a militarily significant oil shortage being created by World War II, that exploration began. In 1946, a geologist named Henry Thomas even published a map in *Oil Weekly*, showing a potential pipeline route that looks remarkably similar to the one we know today. Over the next nine years, though, while a number of exploratory wells were drilled; pointing to the existence of several oil and gas fields, none of them proved to be commercially exploitable.[4]

The future started to look different around Point Barrow after Alaska became the nation's largest state, on August 26, 1958. In an effort to make the new state economically viable, Congress agreed to give the massive new state—with an area of roughly 375 million acres, or more than twice as much land as Texas—the right to choose 102 million acres of that land from anywhere in the public domain. Through its statehood act, Alaska was also given 90 percent of the mineral revenues from an additional 92.4 million acres of Alaska land that were held in various federal reserves.[5]

The state began to choose tracts of land, but in 1961, four Native villages filed protests to the state selections. In the next few years, Alaskan natives blanketed the state with land claims, and by the middle of 1968, approximately 337 million of Alaska's 375 million acres were claimed by one or more native groups. The natives were not the only ones with plans for the land, and the natives' concerns were not exactly given a high priority by most of the state's political leaders, but when the state started planning to sell oil and gas leases on land that natives wanted on the North Slope, the natives protested, as well as banding together politically, forming groups such as the Arctic Slope Native Association and the Alaskan Federation of Natives.

The U.S. Department of the Interior had trust obligations to Alaska natives, although any number of secretaries of the interior had established a pattern of forgetting or ignoring the law in such matters. In 1966, however, Interior Secretary Stewart Udall took the radical step of paying attention to his legal trust responsibilities. He put a moratorium on the state's selection of lands until the native claims were settled, and the Bureau of Land Management, which is part of the Interior Department, halted the sales. Most of the powerful political leaders in the state, by contrast, had a passionate belief in their god-given right to do things the old way. The state's governor at the time, Walter Hickel, an ardent development advocate, held the sale anyway, filing suit against Udall, and condemning the federal action as illegal.[6]

For the most part, investors proceeded to spend money as though they were sure the state and petroleum companies would prevail over the natives in the end. Among the leading companies was a name that would come up again later, namely British Petroleum. By 1966, though, British Petroleum had drilled only seven wells, all of them disappointing. The discovery that changed the future of the state came in January 1968, when Atlantic-Richfield (ARCO) and Humble (later to become Exxon) announced that they had found significant quantities of oil and natural gas at Prudhoe Bay. A few months later, on July 18, ARCO announced that they estimated the find to be 9.6 billion barrels—one of the largest fields ever discovered.[7]

Particularly by an Alaskan scale of miles, Prudhoe Bay was not far from Naval Petroleum Reserve No. 4, both of which were located along the state's northern edge. Both, however, were hundreds of miles away from the nearest port that could actually be used by oil tankers. The 1968 petroleum discoveries provided important incentives to do something about that fact, and by February 1969, officials from British Petroleum, Humble, and ARCO announced plans for a 48-inch, 800-mile-long trans-Alaska pipeline—basically following the same route as the 1946 *Oil Weekly* map—with an estimated cost of about $900 million. Mobil, Phillips, Union of California, Amerada-Hess, and Home Oil companies joined in the venture, and on June 6, 1969, the Trans Alaska Pipeline System (TAPS) applied to the Interior Department for permission to build the pipeline.[8]

There were still those minor annoyances involving the federal moratorium and the legal rights of Alaska natives, but major oil companies had decades of experience with the "flexibility" with which federal laws seemed to be enforced, at least when it came to companies as big and powerful as theirs. The oil companies were so sure they would receive permission to build the pipeline that they invited bids to supply the pipe before they even bothered to *request* the permits to use that pipe, and they actually signed a contract for the pipe during the same month in which they filed the application—June 1969. Pipe deliveries started three months later, years before the permits were obtained.

In the same month as the pipe started to arrive, the state held a lease sale on land around Prudhoe Bay. The average bid came in at $2,180 per acre, bringing the state more than $900 million in income. One bid from the Amerada-Hess-Getty group was for $28,233 an acre, beating the previous all-time record of $27,400 an acre for a single lease, off the shores of Louisiana. This one bid brought more than five times as much income to the state than had its three previous lease sales on the North Slope, which had added up to a total of only $12 million, at an average of a little more than $12.00 an acre.[9]

Part of the reason the oil companies would have bid so much, of course, had to do with the discovery of a massive oil field, but another factor in the oil companies' confidence may have been that an election had been held in 1968. When he took office in January of 1969—a month before the plans for the pipeline had been announced—President Nixon had replaced Secretary Udall with the very man who had sued him—Alaska's former governor, Wally Hickel. Secretary Hickel set up a task force to oversee North Slope oil development, naming Undersecretary Russell Train as its chair. President Nixon expanded this task force to include conservation, industry, and other government agency representatives, asking for a report by September 15, 1969.[10]

High-level government officials usually know how to conduct a study without putting too many roadblocks in front of financially powerful development interests. Over a month before the deadline for submitting the report, Secretary Hickel lifted the land freeze, allowing the building of the first major segment of a road to serve the pipeline. In September, right on time, the task force submitted its preliminary report to the president. It was primarily a status report on the application, but it also noted some problems that didn't actually have solutions at the time, including effects on permafrost, wildlife, and native land claims. The next month, however, Hickel asked Congress to remove the freeze for the entire project, promising that he wouldn't sanction the project until the problems were solved. After initial objections, Congress approved, although one of the reasons for the approval may have been the pending passage of the National Environmental Policy Act (NEPA), which would require a detailed public statement of the environmental impact of any "major federal action significantly affecting the quality of the human environment."[11]

In the meantime, there was that minor matter of the lawsuits that refused to go away. Secretary Hickel was ready to issue the permit for the first phase of the project when native groups and a number of environmental organizations filed suit to stop him. On April 13, 1970, just nine days before the first Earth Day, a federal Judge issued an injunction that did just that. By August, TAPS had reorganized, calling itself the Alyeska Pipeline Service Co. and starting to incorporate environmental awareness into its public image and policies—as for example in ads that promised double-hulled tankers and the safest tanker fleet in the world.

In January 1971, the Department of the Interior released its draft environmental impact statement (EIS), as required under NEPA, and just a few days later, Rogers Morton replaced

Hickel as secretary of the interior. The impact statement briefly considered alternate sources of energy, other pipeline routes, and other modes of transportation, but its main conclusion was that "a pipeline across Alaska would be the least environmentally destructive and most economically advantageous method of transporting the oil deemed essential to meet the nation's rapidly expanding energy requirements." That view was not universally shared; hearings on the impact statement amassed 12,000 pages of testimony, mostly critical, showing widespread opposition to the pipeline, both in the public and in Congress.[12]

Oil representatives, however, ultimately decided to negotiate with rather than to fight one of the most important opposition groups, namely Alaska natives. With the oil industry's encouragement, Congress passed the Alaska Native Claims Settlement Act (ANCSA) in 1971, providing for Alaskan natives to receive 44 million acres of land, along with cash payments totaling almost $1 billion, in settlement for their land claims. Lest there be any confusion about the pipeline, however, ANCSA also specified that, if the secretary of the interior wanted to set aside a pipeline transportation and utility corridor, neither the state of Alaska nor any native groups could select lands within that corridor. In addition, half of ANCSA's cash settlement was to be paid from royalties on oil production, giving the natives an additional incentive to support the rapid completion of the pipeline.

On March 20, 1972, when the final EIS came out, it contained warnings about potential damage to the environment and stated frankly that there was not enough information to provide accurate estimates of many of the project's impacts. The EIS also acknowledged that, in the case of a major tanker accident, most of the oil would end up in the marine environment, although the document went on to editorialize that the national interest in "reducing dependency on foreign oil" virtually demanded that an oil pipeline be built across Alaska, wholly under American control, as soon as possible. The final EIS also argued that an Alaska pipeline would improve the U.S. international balance of payments, create badly needed jobs for Alaskans, and produce essential income for the state of Alaska. On May 11, Interior Secretary Morton announced his decision to grant the construction permits.

That was not the end of the battles, but with oil companies shutting down gas stations across the United States, environmental groups found it increasingly difficult to get Congressional support for the idea of leaving the oil in the ground. By mid-1973, the question was no longer whether to build a pipeline, but where. The industry favored the idea of building a Trans-Alaska pipeline to Valdez, roughly following the route that Thomas had sketched out in *Oil Weekly* more than a quarter of a century earlier, bringing the oil to the lower 48 states with tankers. A number of environmental groups, on the other hand, preferred a route through Canada, sending the oil directly to the states of the upper Midwest. Arguments in favor of the Canadian route stressed that it would go through more environmentally stable terrain, avoid the dangers of tanker traffic, and bring the oil much closer to eastern seaboard markets, all while avoiding the use of oil tankers (which are notoriously vulnerable to enemy attacks) and hence actually being more secure, from a defense perspective—a position that the Department of Defense supported.

President Nixon, however, chose to frame the pipeline, and proposals for expanded offshore drilling, as "domestic solutions" to the energy crisis. Most observers agreed that the Canadian route would cost more, at least in the short run, while taking longer to complete, but in terms of public debate, the more important argument seemed to be that the route through Canada would not be "all-American." In the broader public, meanwhile, panic was building over the prospect of oil shortages. On July 17, 1973, the Senate passed what had become known as the Gravel Amendment to the pipeline authorization bill, declaring that the Department of the Interior had fulfilled all the requirements of NEPA, releasing the pipeline from further legal delay. The final vote was by the smallest margin possible—a tie, which forced Vice President Agnew to cast the deciding vote—and just about two weeks later, the same amendment was passed by the House, then signed by President Nixon.

The OPEC oil embargo provided a final impetus for pushing forward on the construction of the pipeline. On November 8, about three weeks after the start of the embargo, Nixon delivered his speech on the energy crisis, re-emphasizing the importance of Alaskan oil. The Trans-Alaskan Pipeline Authorization Act was passed just four days later, this time by an overwhelming margin, banning any further environmental reviews and restricting further legal challenges to the act's constitutionality. On April 29, 1974, shortly after the end of the embargo, Alyeska began work on the remaining stretch of the haul road north, stretching from the Yukon River to the intended terminus of the pipeline in Deadhorse, Alaska.

The Other Frontier

The other major emphasis of President Nixon's "Project Independence" was to seek a dramatic expansion of offshore oil drilling in the United States, particularly in so-called frontier areas, or those outside of the Gulf of Mexico. That effort, however, was complicated by something else that took place in January 1969—the same month in which President Nixon was inaugurated—the Santa Barbara oil spill.

Opposition to offshore oil development was not exactly a new phenomenon in California. Although the first "offshore" drilling had occurred in Summerland, California, in 1898, the first offshore oil controversy broke out just a few miles away, the very next year. When an oil company began to construct an oil derrick off the shores of Montecito—the highly affluent community that occupies the few miles of coastline between Summerland and Santa Barbara—a local mob, described approvingly on page 1 of the *Santa Barbara Morning Press* the next day as "a party of the best known society men of Santa Barbara, armed to meet any resistance," attacked the rig and tore it down. The local "society men" seem not to have suffered any noteworthy legal repercussions from their actions, despite having been so well known, but oil companies did. The graphic expression of local attitudes was apparently effective in blocking further drilling along that stretch of coastline for decades to come.[13]

The first federal lease sale in the Pacific would not occur until 1963, when it moved forward in the face of local opposition, and sales were also held in 1965, 1966, and 1968. On January 29, 1969, however, the spill in the Santa Barbara Channel effectively ended the ability of the federal government to force Californians to accept new leases. The spill was a "radicalizing" experience for even the most conservative and affluent citizens of the Santa Barbara region—although at least a few of them might have been descendants of the earlier "party of the best known society men" of next-door Montecito. The spill was also radicalizing for many other Americans, who watched the ineffectiveness of response and clean-up efforts on televisions across the country. The Santa Barbara spill has since been credited with providing the inspiration for the National Environmental Policy Act, Earth Day, and much of the subsequent environmental movement in the United States.[14]

By the time work had begun on the pipeline road in Alaska, President Nixon had other problems on his mind—the scandal that some have seen as the most important one to shake the White House since the day of Teapot Dome, and that others see as having been even more significant—the aftermath of the break-in at Democratic National Committee headquarters, in an office complex called Watergate. Public interest in the scandal had been rising along with energy prices during most of 1973, particularly as the Senate held Watergate hearings during the summer of that year. On October 20—two days after the start of the OPEC embargo—Nixon dismissed independent special prosecutor Archibald Cox and accepted the resignations of Attorney General Elliot Richardson and Deputy Attorney General William Ruckelshaus, in what came to be remembered as "the Saturday night massacre." By March 1, 1974—two days after the start of construction of the new section of the Alaska pipeline's service road—a grand jury indicted many of the president's closest former aides, who came to be known as the "Watergate Seven." Another five months later, on August 9, 1974, President Nixon became the first president in American history to resign from office.[15]

For policies regarding offshore drilling, it made little difference when Nixon's recently appointed vice president, Gerald Ford, took on the presidency. In all but one important respect, it also made relatively little difference when Ford was defeated by the man who became president in January 1977, Jimmy Carter. The one exception was that President Carter and his appointees saw the use of negotiation and compromise as being part of good government, rather than as a sign of weakness.

It was also a new approach. For the most part, the Department of the Interior's approach to holding lease sales on the Outer Continental Shelf had long been one that indicated little concern about the objections of affected regions. In 1975, the

Interior Department had offered more acres for lease off the shores of Southern California than had been offered in all of its previous lease sales, combined. During the next two years, the department held the first OCS lease sales off the East Coast and in Alaska, along with the largest annual acreage offerings to date in the Gulf. Still, while the stated intent was to open up new areas for oil and gas exploration, one of the major effects was instead to open up a much higher level of opposition than the agency had ever encountered before—making it possible, at the same time, for opponents from different regions to join forces in opposition to any such expansion.[16]

On the east coast, the potential for vastly increased offshore leasing led to the formation of the Mid-Atlantic Governors Coastal Resource Council, which sent representatives to examine the effects of developments in the Gulf. Those representatives were apparently appalled by what they saw. The governors passed a resolution, stating their concern and calling on Congress and the president to undertake "with the full and active involvement of the States ... a detailed and accurate analysis of the onshore impacts of the program." By 1976, the second and last full year of the Ford presidency, Alaska went to court to try to stop a lease sale in the Gulf of Alaska. The two states in which the opposition was strongest, however, were Florida and, especially, California.[17]

In California, the key event was a proposed lease sale that started with a "call for nominations" just a few months after the start of the Carter presidency, in November of 1977. Unlike previous sales, which were concentrated in one geographic region, this one called for nominations of tracts that along the majority of the state's coastline, from Santa Barbara, along the south-central coast, all the way north to the Oregon state line. A new coalition, representing a score of interest groups, was formed to oppose the sale, and the Interior Department was soon facing the stiffest opposition its offshore leasing program had ever encountered.

The opposition, in combination with the Carter administration's willingness to engage in dialog, may have helped Congress to pass legislation that it had been debating for the past four years—the Outer Continental Shelf Lands Act *Amendments* of 1978, or OCSLAA—the first significant change in the laws governing offshore oil development in a quarter of a century. The Legislative History of the Act, also known as P.L. 95-372, expresses Congressional concern that OCS leasing and regulation were proving to be a closed decision process, involving just industry and the top-level officials of the Department of the Interior. One of the stated purposes of OCSLAA, by contrast, was to open the decision-making process to a wider audience, thereby increasing public confidence—all while expediting the development of the resources of the Outer Continental Shelf.[18]

The new law was scarcely the end of public controversy; the states of California and Alaska challenged one of the Interior Department's first major actions under the Outer Continental Shelf Lands Act Amendments, namely the agency's 1979 announcement of the first five-year schedule for leasing, which was required under OCSLAA and which was announced by the department in 1979. The states, moreover, eventually won in the Washington, D.C., District Court. When the protests extended into 1980, Carter's secretary of the interior, Cecil Andrus, withdrew much of the area that had been scheduled for leasing off the coast of California, as well as canceling a number of controversial tracts. Later in that same year of 1980, however, Ronald Reagan would be elected as the new president, and his would be an administration with no interest whatsoever in idea of slowing down the leasing process to respond to local concerns. As if to underscore that point, the new administration's secretary of the interior would be none other than James G. Watt—someone who would later be described by the chief environment counsel at the House Energy and Commerce Committee at the time as one of the two most "intensely controversial and blatantly anti-environmental political appointees" in American history.[19]

If the central theme of the late 1970s had been compromise, the central message of the new administration was that the time for compromise was over. In early 1981, shortly after the Reagan administration was sworn in, the new secretary of the interior, James Watt, in his first public policy statement, announced he was reversing a negotiated settlement between former secretary of the interior Andrus and California, dramatically increasing leasing offerings off of California. California, joined by nine other coastal states and a number of cities and counties, had other ideas.

8 To Know Us Is to Love Us?

In 1990—ten years after the start of the policies that had been put in place by James Watt, and almost exactly twenty years before the blowout of the *Deepwater Horizon*—the two of us started a small pilot study for the very agency that James Watt had shaped to implement his policies, namely the U.S. Minerals Management Service. At the time, one of us had served for several years on that agency's own Scientific Advisory Committee, and the other was familiar with the agency from having lived in southern Louisiana and from having served on an independent review committee established by the National Academy of Sciences/National Research Council. The agency, meanwhile, had experienced remarkably little progress in implementing the expansion of leasing to "frontier" areas that James Watt and his successors had championed.

This may have been one of the reasons that MMS decided to support our study—one of the smallest in the history of its Environmental Studies Program. As the agency's top officials put it at the time, they were quite interested in a study of what they consistently called public "risk perceptions" regarding offshore oil development. The reason for that, in turn, was that those officials shared a perception of their own. As that view was summarized by one long-time observer of the nation's offshore oil policies, "Louisiana has offshore development, and they love us. California doesn't have offshore development, and they hate us. Therefore, to know us is to love us."

Our study led to a series of peer-reviewed articles and a book, as well as to an official report to the MMS itself. Both in our report to the agency and in the publications that were intended for a wider audience, we tried hard to make it clear that loyalty would best be measured by candor, rather than by adherence to a party line, and that true friends are those who, even if they use the kindest words they can muster, will still tell you honestly what other people are saying about you. This, however, did not seem to increase the popularity of our findings, at least within the upper ranks of the Minerals Management Service. The reason may well have been that, to put things simply, our findings showed that the strange state, for once, was not California, but Louisiana.

On a date that happened to involve the Ides of March 1989, an offshore oil lease sale was held in New Orleans, for tracts off the coast of Louisiana. There were no protestors, no state opposition, and few comments of any sort. At the sale, the Department of the Interior uneventfully leased 2.8 million acres off Louisiana to oil and gas companies, bringing over $380 million to the federal treasury. In the words of one of the persons we talked to in doing our study, this lease sale showed once again the predominant attitude in Louisiana: "Good luck—hope you find some oil—and if you do, send us a check."

A few months later, by contrast, a presidential task force would be sent to Tallahassee and Key West, Florida, to hold hearings on offshore oil leases that had been proposed off the southwestern coast of Florida. The agency was pushing hard to hold the leases, despite years of opposition from two consecutive Florida governors, Bob Graham and Bob Martinez. The crowd in Key West was estimated at between one and two thousand, and a somewhat smaller crowd in Tallahassee was equally adamant. With the exception of brave spokespersons from the oil industry, almost no one in either crowd had anything positive to say about the idea of permitting offshore oil development off the coast of southwest Florida.[1]

The opposition in Florida was almost as intense as the kind of reception that offshore oil proposals were inspiring in California, and it was clear that none of the citizens' reactions could actually be blamed on a small handful of ignorant malcontents. Instead, as a number of Louisiana residents told us, what we needed to explain to the federal government was that a set of special factors made Louisiana a congenial location for oil development—and that the absence of such factors would provide reasons to expect the same kinds of oil activities to be much more controversial around most of the rest of the nation's coastline.

The factors that make Louisiana unusual start with the history of the offshore oil industry: Coastal Louisiana is the region where the offshore oil industry was invented and developed. Particularly from the 1950s through the early 1980s, OCS development led to unparalleled opportunities for local entrepreneurs, as well as for employment. Eventually, the export of offshore oil technology became a major element of the region's economy. Major portions of the first platforms in the Santa Barbara channel, for example, were built not in California, but in Louisiana. Yet it is also important to realize that oil development in coastal Louisiana started in the 1930s and 1940s—decades before the first Earth Day, or for that matter, the emergence of most other environmental concerns. If the initial push offshore had taken place in a different part of the country, but at around the same time, the results might well have been similar. By today's standards, for example, the initial drilling along the California coast at the beginning of the twentieth century would scarcely be considered an environmental success story.

Because of the early start, moreover, offshore development in Louisiana was already in place before a number of potentially competing uses were established. In contrast to proposals for OCS development in regions that are under heavy fishing pressure already, for example, oil development off the shores of Louisiana got started at a time when the state had little prior tradition of offshore fisheries. Shrimp, the most significant of the commercial species in the Louisiana Gulf in recent decades, were thought to be present only in the estuaries, not even being known to be available in the open Gulf until the 1950s. In essence, both of the major offshore activities in Louisiana—oil and gas development, and the harvesting of fish and other types of renewable resources—essentially grew up together. As noted earlier, oil exploration activities even used fishing boats that were leased by the oil companies. Morgan City, where the first truly "offshore" drilling effort took place, still has an annual Shrimp and Petroleum Festival—the official symbol of which is a shrimp in a hard hat, wrapped around an oil derrick.[2]

Rather than seeming like a foreign activity, accordingly, OCS development in Louisiana has always been a local affair. In general, it has been in the Gulf of Mexico that offshore technology has evolved—starting as a gradual extension of earlier, onshore activities, and often focusing on developing local solutions to technological problems—leading to an understandable pride in the region's accomplishments. The evolution of offshore technology was paralleled by the similarly gradual emergence of support services and by altered forms of work scheduling (seven days on and seven days off, etc.), which were adaptations to the logistical problems associated with operating at remote sites. While the Northern waters of the Gulf of Mexico have become the most heavily developed offshore area in the world, the development has taken place one step at time. Clearly, such a history is very different from proposals for development in potential "frontier" regions, where proposals have long called for massive, technologically sophisticated, and capital-intensive developments that would land on the scene almost as suddenly as an alien invasion.

Louisiana, however, is also *physically* different from most coastal regions of the United States, both onshore and off. Onshore, Louisiana has long been the home to the most extensive system of coastal marshes in all of the United States. One consequence is that—unlike most coastal states, where waterfront property is highly prized and densely populated—Louisiana's residents rarely live anywhere near the coast. Most of the coast is lined with a broad and virtually impenetrable band of marsh—well-suited for fish and wildlife, but not so easy for humans to inhabit. In many places it is effectively impossible to get within *ten miles* of the coast by road, and local residents' descriptions of their state's coastal regions are more likely to involve mosquitoes and alligators than spectacular visual imagery. As a resident of southern Louisiana once explained, "The Gulf is only about 15 miles south of here, but there are probably more people in this town who've seen the Gulf from *Florida* than who've seen it from anyplace in Louisiana."

In most coastal states of the United States, the situation is virtually the opposite: Most of the population lives on or near the coast, and the coast is readily accessible by road. In most of those states, the coast is seen as a valuable public resource, a thing of beauty, and a source of popular recreation—partly because the coast can actually be seen by so many people, so often. Even in Alaska, while there are relatively few miles of highways along the coastline, or for that matter anywhere else, a significant fraction of the state's official highway system is literally *on* coastal waters, in the form of ferries that ply what is officially known as the Alaska Marine Highway. At least by Alaska's standards, the coast is thus relatively accessible, and it is viewed as a resource and an important recreational feature. By contrast, as one Louisiana respondent noted in the era before the BP blowout, "We don't have beaches, we have estuaries. You couldn't go walking along the estuary and find a dead bird even if there was one."

While it is difficult for humans to reach the Louisiana coast from land, however, access to the land from the water is considerably simplified by the numerous bayous of the region. Unlike many coastal areas in the United States, Louisiana is characterized by an abundance of waterways that can provide marine access and harbor space. Anyone who drives along the rivers and bayous of southern Louisiana is likely to see a continuing parade of piers—some with fishing boats, others with a broad range of oil-service craft—with many more miles of bayous offering waterfront locations that could be turned into harbor space for additional vessels at relatively minimal expense.

Offshore, moreover, the topography of Louisiana's continental shelf presents very different conditions from those that are found in areas where OCS development has been most contentious. In most areas of the country, but particularly along the Pacific coast, there is little in the way of a "continental shelf," since the ocean bottom drops off quickly, even dramatically. In the Gulf, by contrast, production platforms are in place well over 100 miles offshore, and in some areas, the slope is as gradual as one or two feet per mile. The breadth of Louisiana's continental shelf reduces the likelihood of use conflicts, while simultaneously reducing the number of problems created by any given obstacle. Even if a fishing boat needs to make a quarter-mile detour around an oil rig, there will be little impact on that ship's ability to keep its nets in contact with the sea floor. In areas with steeper ocean bottoms, by contrast, less area is available at any given depth or contour line, and the presence of an oil platform can require major adjustments for any boats that try to "fish around" such a obstacle.

Equally important, however, the muddy discharges from the Mississippi and Atchafalaya rivers mean that the sea floor of the Gulf of Mexico is mainly covered in silt—as well as meaning that oil rigs can provide a significant advantage for fishing operations. Certain species of commercially important fish can survive only in the kinds of habitat known collectively as *hard substrate*—rocky bottoms, reefs, rock outcroppings, and the like. Particularly in the areas off Louisiana where oil development activities have been the most intense, natural outcroppings of this sort are so rare that oil-related structures now make up more than a quarter of all hard substrate. In effect, the oil rigs thus serve as artificial reefs, concentrating the fish populations, and it is quite common to see fishermen congregating near oil rigs as well.[3]

That same silt that makes the artificial reefs important also means that, in contrast to many coastal areas in the United States, when Louisiana residents think of coastal waters, they don't think of them as clear. Local commercial divers, only half-jokingly, refer to work in the bayous and canals as "mud diving." For a diver who needs to cut out a rope that has been sucked into the propeller of an oil-field supply vessel along the Gulf Intracoastal Waterway in Louisiana, it doesn't matter if the diver's eyes are open, closed, or literally blind—the water is opaque.

Louisiana, finally, was also *socially* distinctive, particularly if we compare Louisiana at the time when offshore development started to expand against the conditions that are found in most coastal regions of the United States today. Particularly notable are the education levels and the extractive orientation toward the local environment at the time when the expansion of offshore oil drilling took place, and the patterns of social contacts that had come to characterize the region by the time when Ronald Reagan was elected.

Studies tend to find such broad support for environmental protection that educational levels are among the few sociodemographic predictors that show any significant correlations with environmental awareness and concern; better-educated individuals in the United States generally express higher levels of environmental concern. Thus it may be significant that, particularly in the 1930s and 1940s, coastal Louisiana had some of the lowest educational levels in the country. For example, in St. Mary parish, the scene of initial OCS activity, only 47.2 percent of the adult population had five or more years of education in 1940, and only 12.2 percent had graduated from high school. Other rural areas of southern Louisiana had similarly low educational levels. By way of comparison, well over 75 percent of the adults in the United States today have a high school education or more.[4]

At the time of initial OCS development, moreover, the existing economy in coastal Louisiana was dominated largely by *extractive industries*—those that, like oil development, involved the extraction of raw materials from nature. As a general rule of thumb, persons who are involved in extractive activities will be less likely to object to new extractive industries than will persons in manufacturing or service industries, and they will be far less likely to object than will persons whose livelihood depends on the maintenance of high environmental quality. Extractive industries, however, have been shrinking rapidly as a proportion of the U.S. labor force, dropping by roughly two-thirds, in proportionate terms, since 1920—as sharp a decline, on average, as the better-known decline in the proportion of the workforce engaged in farming. In most coastal regions of the United States today, the economy is far more dependent on the amenity values of the coast than on its extractive values, and the likelihood of finding support from extractive workers can be expected to continue to decline.[5]

By the time when James Watt wanted to expand offshore drilling, finally, southern Louisiana was also a place with a strong and unusual *social multiplier effect*. Studies often note that the economic impact of an industry will include not just the wages that are brought into an area, but also the fact that when those wages are spent in local stores and businesses, they have a further, *economic multiplier* effect. In comparable ways, a given person's attitudes toward an industry are likely to be affected not just by whether that person works for the industry in question, but by whether her friends or neighbors do. It's one thing to decide whether to support or oppose a new industry, in the abstract, but it's another matter entirely to make such a decision about an industry that employs your next-door neighbor, or even your neighbor's second cousin. Given the historical, biophysical, and other social factors summarized above, the "average" resident of coastal Louisiana in the 1940s would be expected to have known friends and neighbors who were employed in the oil industry, and by the 1980s or 1990s, it would have become virtually impossible to live in most communities of southern Louisiana without knowing someone who was so employed. In regions that are dependent on amenity values of the coastline, by contrast, today's residents are far more likely to have relatives, friends, or other acquaintances whose livelihoods could be devastated if the beach were suddenly to be trashed by tar balls.

All in all, our recommendation to MMS was that the agency should not expect public opposition to offshore drilling to be changed to any significant degree by what the agency called "public education" programs. The people who lived in coastal Louisiana really were as supportive of offshore drilling as they seemed to be, but many of the key reasons were ones that had little relevance to most other stretches of the nation's coastline, with the possible exception of Alaska. The people who lived along other stretches of the coastline, meanwhile, were already "educated" about offshore drilling. Their opposition was not based on ignorance, but on what were sometimes impressively clear understandings about what an expanded drilling program might mean for what they cared about most. To know the agency better, we reported, was not necessarily to love it.

Our final report to MMS seems to have been significantly less popular at higher levels of the agency than was the more abstract idea of doing a report on "public risk perceptions." It disappeared more or less without a trace. Instead, the agency continued to press forward in the effort to implement the policy changes first put into place by James Watt.

There were actually two main thrusts to those policy changes. One was to open up more areas for leasing outside of its established range along the central and western portions of the Gulf of Mexico. The other was to expand the areas being leased, period. The first of those intended changes ran into roadblocks that continued to neutralize Secretary Watt's goals for decades to come. The second, which received far less scrutiny, went forward largely as planned, and with consequences that have received remarkably little attention since that time.

Up until the Reagan administration, established procedures within the Department of the Interior called for the Bureau of Land Management (BLM) and the U.S. Geological Survey (USGS) to begin by conducting a resource evaluation of a broad offshore area, after which an invitation to nominate tracts within a given area was published in the *Federal Register*. The tracts would be limited to three-mile-square "blocks," and the selection of the offshore tracts to be offered would be based on the indication of interest by industry, sometimes on expressions of concern by environmental groups and/or affected states, and on the resource assessment conducted by USGS.

When Watt came to office, he combined all offshore activities in a new agency with a name that left no doubt about its central focus—the Minerals Management Service—and he threw away the old procedures, replacing them with what MMS called *area-wide leasing*. In simple terms, that meant opening the leasing process to entire planning areas (e.g., the western Gulf of Mexico), rather than continuing the much more limited offerings of specific of tracts, as had been the case since the first Federal lease sale in 1954. The proposal for area-wide leasing met with almost unanimous opposition from coastal states, including Louisiana. Even after the controversial James Watt was replaced by William P. Clark, Jr.—a personal friend of President Reagan's, and someone who was far less outspoken than Watt—the policies of the Department of the Interior regarding offshore oil development showed little sign of compromise or change. One of the lower-level officials we knew at the time offered the most concise summary we heard of the agency's position—he was convinced that the opposition would quickly fade away, he said, because the idea of area-wide leasing "could only be opposed by the ignorant or the criminally insane."

His prediction would be tested almost repeatedly over the next several years, and the results were less than impressive. In fact, it should perhaps not have come as a surprise that, if even the state of Louisiana opposed the change, other states would, as well—and none of them seemed to be charmed by having their intelligence or sanity questioned, or even by hearing the somewhat more polite versions of his charge, repeated frequently, that the opposition "just" reflected ignorance, selfishness, or irrationality, rather than reflecting legitimate concerns on the part of sensible citizens.

These, clearly, are not the kinds of descriptions that seem to indicate an eagerness to understand one's opponents better. Instead, perhaps their major effect was to contribute to a phenomenon that the two of us have called a "spiral of stereotypes": When persons on different sides of an issue stop talking to one another, but persist in talking *about* one another—particularly if they are characterizing the concerns of the other side as being ill-informed, self-serving, or irrational—the net effect can be to increase further the amount of polarization that was already present. Perhaps it should not come as a surprise, then, that residents of other coastal regions seem to have had equally little interest in developing warmer relationships with representatives of MMS and the oil industry. As noted above, when Louisiana citizens think of oil industry representatives, they are thinking of their friends and neighbors, but by contrast, in the words of one Californian we interviewed in the 1990s, "When I think of the oil industry, I think of fat, pushy Texans in pointy-toed boots."[6]

One key effect of the strong insistence on expanding the offshore drilling program, and leaving behind the spirit of compromise from the Carter years, was thus a political version of Newton's third law of motion. The insistence on strong federal action led to an equal and opposite reaction. At least as important, though, not only did the new program increase the opposition to drilling along the Outer Continental Shelf—instead, it also served to unite states that had previously differed significantly in their views. Again, California went to court, this time joined by Alaska, Washington, Oregon, and Florida. In this case, though, the states were not successful in their litigation, and the following year Watt initiated area-wide leasing, at least in the Gulf, drastically increasing the acreage offered for lease.[7]

While Watt managed to expand the area for lease in the Gulf, however, he was far less successful in other regions. Instead, responding to newly unified and much stronger patterns of opposition, Congress started inserting prohibitions into the annual appropriations for the Department of the Interior, forbidding the expenditure of funds for leasing activities, particularly off the east and west coasts. In effect, this canceled out the decision-making process within the department, shutting down any expansion of Outer Continental Shelf leasing, a year at a time, outside of the western and central Gulf of Mexico. In spite of these warning signs, Watt and his successors continued to push for such expansions—and the Congressional moratoria became an annual event. Gradually, the prohibited areas grew in size and geographic distribution, to the point that practically the entire Outer Continental Shelf was effectively closed off to the very agency that had been created to exploit it.[8]

In the spring of 1989, the newly elected George H. W. Bush inherited the Outer Continental Shelf gridlock generated by his predecessors. There had been only two Outer Continental Shelf sales outside of the Gulf of Mexico in the past five years—both of them off the north coast of Alaska—and Congress showed no inclination to relinquish its control over the leasing process.

The most intensive controversy at that time when he took office centered on three proposed sales in California and Florida. One of the first actions he took as president, early in 1989, was to announce that he was postponing all three proposed sales, submitting to a cabinet-level presidential task force the question of whether the sales should go forward, and asking the National Academy of Sciences to evaluate the adequacy of the available information concerning the potential impacts of the sales.

Both the task force and the academy hearings on the subject resulted in dramatic forms of public involvement. As one observer within the Minerals Management Service told us at the time, though, the new president's decision to involve the

National Academy seemed to reflect a belief that was almost universal within MMS, or at least among those who held high positions in the agency. The party-line view, which we ourselves certainly heard often enough, was that all of the scientific questions about the impacts of offshore development had already been answered, meaning that any remaining opposition was based merely on disagreement with the administration's policies. As he put it, "They figured that the Academy would tell people. 'You're wrong, the science is fine, so why don't you just shut up?'"[9]

As he went on to add, however, "I guess they figured wrong."

By late 1989, the response to the president's question from the National Research Council, the working arm of the National Academy of Sciences, came in the form of a report—one that contained few of the usual academic hedge words. In polite but uncommonly clear language, the report made it clear that, much as critics had charged, the available scientific information truly was inadequate for informed decision making—and for all three sales. Shortly thereafter, when the Cabinet-level task force presented its findings to the president, its report offered a number of options, but none of them included going ahead as planned. Several weeks later, Bush canceled the three sales and imposed a presidential moratorium on lease sales in these three areas and on much of the east and west coasts of the United States. Presidential moratoria remained in place for the next eighteen years, until the final year of the administration of George W. Bush, the son of the first President Bush, who announced the change in the run-up to the 2008 elections.[10]

Along the Gulf coast of Louisiana, by contrast, the effect of Watt's change in policy was to lead to an even more frenzied version of the boom that had already gripped much of the southern tier of the state—at least for the next few years. One of the key cities in the oil region even received hyperbolic attention in the *National Enquirer*, being reported (erroneously) as a place where two out of every thirty residents were millionaires, teenagers sported diamond-studded Cartier watches, and people hopped on Lear jets for lunch hundreds of miles away. The boom was certainly on, but it wasn't quite that spectacular.[11]

Then, unfortunately, came the bust.

One of the little-recognized successes of president Carter's energy policies is that his emphasis on conservation actually achieved significant improvements, including the only significant drop in oil imports during the past half-century. By 1981—the first year of the Reagan presidency, when he and James Watt inaugurated the drive toward more aggressive leasing of offshore lands—U.S. petroleum consumption had actually fallen below the levels that had been reached before the 1973–1974 embargo. Given memories of oil shortages and soaring prices, consumers cut their demand in a variety of small ways; like buying more fuel efficient cars and insulating their houses. Decreases in demand did not show up at first in the price of oil. Prices fluctuated but stayed relatively strong for several years, remaining at $24.51 per barrel as late as December of 1985. By June of 1986, however, the price for the same amount of oil had fallen to $9.39. Once the global prices began their steep dive, local consequences were quick to follow: Unemployment rates in the oil-dependent regions of coastal Louisiana, which had averaged in the 4 to 5 percent range for the previous decade, rose past 20 percent in many areas of the oil patch by the end of 1986.[12]

The collapse was greeted with astonishment as well as anguish. Many of the region's younger residents had never known a time when coastal Louisiana was *not* booming, and at least at first, there seemed to be reason to hope that the sudden plunge would be a short-term aberration. As if to underscore the fact that Louisiana's energy-related bust was not a small or short-term matter, however, *New Orleans Magazine* noted a half-dozen years later, in its January 1992 issue, that of the area's 25 Savings and Loans that had managed to survive for five years after the disastrous price drop that followed December 1985, only 14 "remained untouched by regulators and met minimum federal capital standards" as of the period ending six months later.[13]

At the national level, as well, any effects of expanded leasing on "energy independence" proved to be brief. As indicated by figure 8.1, the drop in oil imports came to an end soon after president Reagan took office. The all-out effort to expand leasing and remove federal constraints on domestic energy production did lead to a small reversal of previous downward trends in U.S. oil production, at least during the first few years of the Reagan presidency, but the change was scarcely overwhelming. Before the end of that presidency, moreover, the realities of geology had asserted themselves once again, and the downward slide resumed—accompanied by a sharp increase in oil imports. By the time President Reagan left office in early 1989, the Unitged States was providing an even lower share of the oil it consumed than had been the case when he first took office.

The Effect of "Area-Wide" Sales

As a reminder, the federal "take" from the exploitation of offshore resources comes from several sources, but the primary "profit" comes from bonuses ("competitive" bids on offshore tracts by oil companies to obtain a lease, which can vary a great deal) and royalties (a percentage of the resources being extracted, at rates established by the Department of the Interior). Perhaps the clearest single change resulting from Watt's efforts to push new policies was a dramatic expansion of the area being offered for lease in the program's "non-frontier" region,

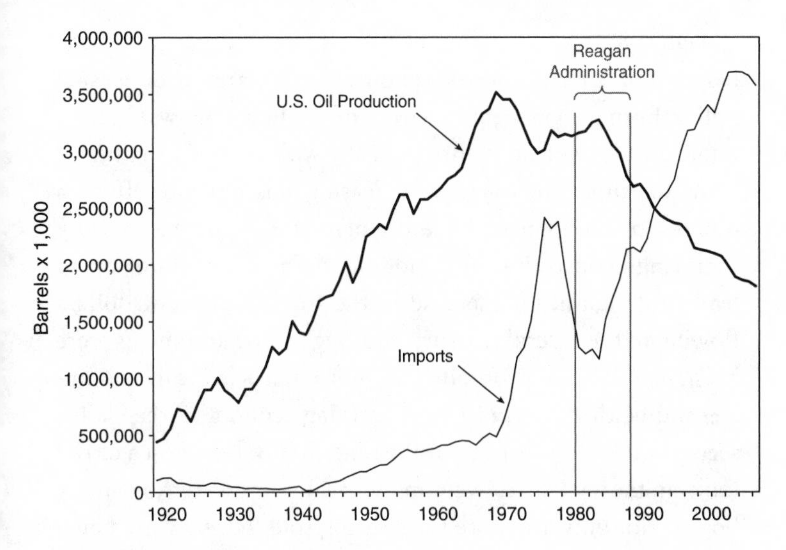

Figure 8.1
U.S. oil production and imports

the central and western Gulf of Mexico. Prior to the start of so-called area-wide leasing, the all-time record lease sale had offered 2,870,344 acres to industry. The first area-wide sale offered more than a dozen times as much territory—37,867,762 acres. While Watt's other policies were largely stopped by the intensity of public opposition, moreover, the move to area-wide leasing was never challenged or reversed.

Given that leasing on the Outer Continental Shelf started in 1954, we now have almost half a century's worth of evidence on the results of that program. Area-wide sales started about half-way through this period, in 1983, creating an *interrupted time series*, or in simpler words, a good opportunity to compare the effects of the Watt policies, as opposed to those that had been in place before he took office. At least two questions can easily be addressed—how many leases were sold annually and how much revenue the federal government collected for us, the taxpayers, on these sales.

Figure 8.2 shows the number of leases sold per year, both before and after the introduction of Watt's area-wide leasing policy. Even a casual examination of the figure shows that the number of leases shot up dramatically with area-wide leasing.[14]

At the time when area-wide leasing was put into effect, a number of economists we know made the argument that, by removing "constraints" on bidding, the new program would lead to a significant increase in the total number of dollars flowing to the federal treasury. Although those arguments were bolstered by the appropriate jargon from neoclassical economics, and neither of us is a card-carrying economist, the claims seemed questionable to us at the time. Today, based on a closer look at the evidence, we can see that they were absolutely bogus. Not only are more tracts being sold, but they are being sold with fewer bids per tract—and at bargain-basement prices.

Figure 8.3 shows the mean number of dollars per acre that we, the taxpayers, have received in accepted bids for

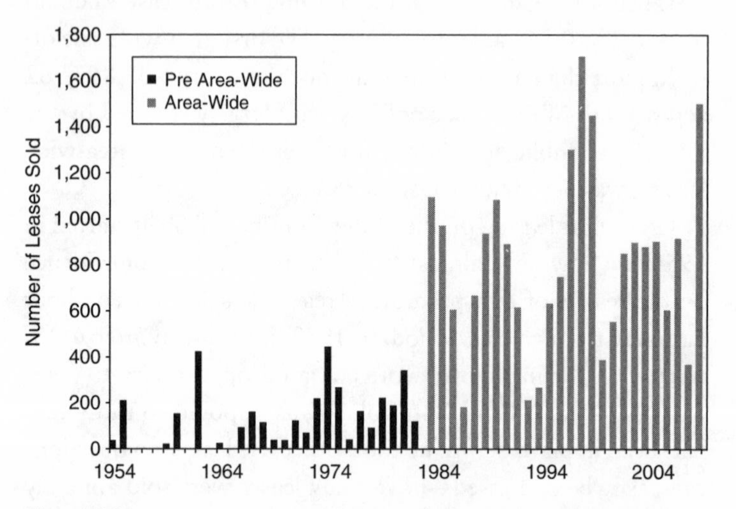

Figure 8.2
Federal offshore leases sold, before and after start of area-wide leasing

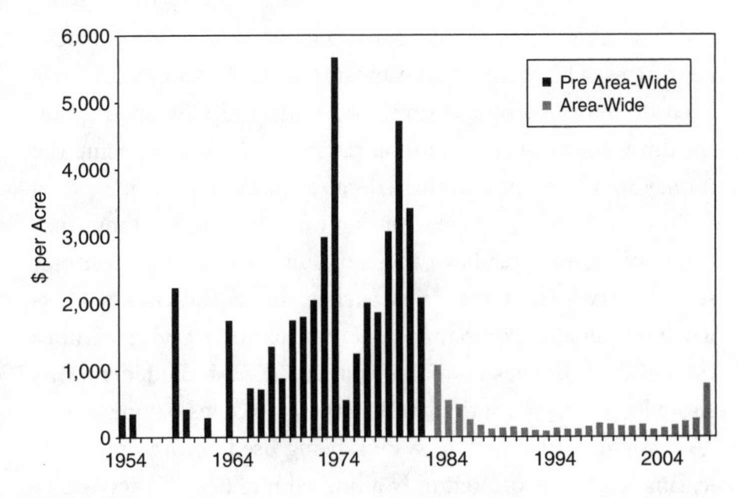

Figure 8.3
Federal income, in dollars per acre, before and after start of area-wide leasing

leases—*without* taking into account the effect of inflation, which would make the drop-off in income even more dramatic. Again, even a casual examination of the figure demonstrates that oil leases have consistently sold for far less federal income, on average, after the start of the Watt-era policy. Despite more than a six-fold expansion in the number of leases sold, the total monetary amounts actually went down, not up. Before area-wide leasing started in 1983, 3,520 leases were sold at an average dollar-per-acre rate of $2,224. From 1983 to 2008, 21,179 leases were sold at an average dollar-per acre rate of just $263. Multiplying the numbers shows that—again, without even taking account of the fact that inflation has taken a huge bite out of the value of the dollar since 1983—oil companies have actually paid almost 30 percent less in *total* bid dollars under Watt's system than under the system that was in place up to 1983.[15]

Yet that's not all. When the Obama administration proposed a tougher set of regulations for offshore oil development, one of the responses from his opponents in Congress was that tougher regulations would drive out the smaller or independent firms. The reality is that most of those firms have already been driven out—by the same policy that rapidly speeded up the draw-down of American oil reserves, all while reducing the money that has come to the federal treasury in return.

In simple terms, it turns out that only the largest of multinational oil companies have the economic resources to contract seismic surveys on large tracts, such as the central Gulf of Mexico, while smaller companies generally do not. At present, more than 400 companies own all or part of at least one federal offshore lease, but more than half of these holdings belong to just 20 companies. Under the system being used before area-wide leasing, each sale offered only a limited number of tracts—giving smaller companies a fighting chance of being able to find out enough about some of those tracts to submit informed bids. This meant in turn that there was enough potential for competition to encourage larger companies to avoid low-ball bids. The move to area-wide leasing meant that the few companies having extensive resources for geotechnical exploration could identify what they considered prime properties, without worrying too much about the possibility that other companies might want the same properties. For all but the biggest companies, Watt's system amounts to a process of tossing a few darts at a huge map of the Gulf. Much the same is true of the government agencies that supposedly protect the interest of the taxpayers: In effect, area-wide leasing sets up a sale situation where the buyers (major oil companies) know the potential value of a given tract, but the seller (the federal government) does not.

Royalty Rates

In addition to the fact that bids to acquire leases dropped significantly after the start of the Watt-era policies, the United States receives a lower royalty rate than almost any other country. Traditionally, offshore leases required that the federal government receive the highest available bonus bids, plus one sixth of the total value of offshore resources extracted (16.66 percent). When Area-wide leasing started in 1983, a number of leases were sold with lower requirements, for just one-eighth of the value, or 12.5 percent. These lower rates were generally offered for leases with greater water depths, and they were described as being an attempt to encourage leasing and exploration in deep-water frontier areas. A dozen years later, the first Congress to be elected after the Republicans took over the Congress in 1994, as part of the "Gingrich revolution," passed what it called the Outer Continental Shelf Deep Water Royalty Relief Act of 1995. The stated purpose of the act was to "encourage production of marginal resources" on tracts in the deep-water areas of the Gulf of Mexico, defined as those in water depths of more than 200 meters, or about 650 feet. In addition to lower royalty rates, this act suspended *all* royalties for five years in a tiered system that allowed lease holders to produce millions of barrels of oil and billions of cubic feet of natural gas without paying *any* royalties to the American taxpayers who owned the resources in question. Needless to say, these terms were appealing to fossil fuel companies, leading to the high number of leases that can be seen from 1996 to 1998 in figure 8.2—a total of 2,840 leases in the three years following the passage of the Deep Water Royalty Relief Act in November 1995.[16]

The results have not been so favorable for American taxpayers. As the Government Accountability Office has noted,

> Based on results of a number of studies, the U.S. federal government receives one of the lowest government takes in the world. Collectively, the results of five studies presented in 2006 by various private sector entities show that the United States receives a lower government take from the production of oil in the Gulf of Mexico than do states—such

as Colorado, Wyoming, Texas, Oklahoma, California, and Louisiana—and many foreign governments.[17]

Tax Incentives

In addition to price breaks on the resource they extract, oil, the industry operates under some of the most generous tax breaks in the United States. Some of these tax policies grew out of foreign policy considerations that are no longer relevant. For example, in an effort to combat Soviet influence in the Middle East during the cold war, the State Department allowed a Saudi Arabian accounting maneuver that classified the royalties that American oil companies paid to foreign governments to be classified instead as "taxes," which entitled the companies to deduct these payments from their U.S. taxes. Proponents argue that these tax breaks offer incentives to discover and produce new sources of oil, but critics such as Sima J. Gandhi, a policy analyst at the Center for American Progress, make a different argument: "We're giving tax breaks to highly profitable companies to do what they would be doing anyway.... That's not an incentive; that's a giveaway."[18]

An assessment by the chief economist of the Treasury Department during the first year of the Obama administration argued that it was time to repeal a set of six "tax preferences" that, according to his calculations, would otherwise cost the Treasury about $30.6 billion—with a "b"—over the next ten years. The new administration also argued for dropping two other sweetheart deals for the industry that would kick in with billions of dollars of additional subsidies, but only if oil prices dropped. Drawing on the earlier assessment by the Congressional Budget Office, the Treasury Department economist offered the following overall assessment:

> The tax subsidies that are currently provided to the oil and gas industry lead to inefficiency by encouraging an over investment of domestic resources in this [oil] industry. In 2005 the Congressional Budget Office (CBO) estimated that the effective marginal tax rate on investment in petroleum and natural gas structures was 9.2 percent. This is well below the average effective marginal tax rate for all asset types (26.3 percent). The size of the distortion ... is quite large ... and removing this distortion would improve overall economic efficiency.[19]

In ordinary English, the oil industry gets so many tax breaks that the net result is to "reduce economic efficiency," or to *hurt* the overall economy, doing so by diverting money away from other industries that might actually make better use of the investment dollars. As noted by a headline the *New York Times* in the aftermath of the BP disaster, however, "As Oil Industry Fights a Tax, It Reaps Subsidies." The article noted that, even as the oil was continuing to gush into the Gulf, during a time of intense public anger toward the oil industry, efforts to end such ill-advised tax breaks were "likely to face fierce opposition in Congress," partly because the oil and natural gas industry had spent $340 million on lobbyists—more than a third of a billion dollars—between 2008 and 2010 alone. The industry seems to have had good reason to invest so heavily in their lobbyists—the snake-oil salesmen of today. As the article went on to note, "only the industry's political muscle is preserving the tax breaks." The lobbyists' arguments usually included the claim that the end of their sweetheart deals would hurt "the economy," but Treasury Department analyses concluded that they would "hurt" only the economics of oil companies. Oil prices and potential profits have been so high, so consistently, that eliminating the subsidies for oil companies "would decrease American output by less than half of one percent," while *improving* the nation's economic performance overall.[20]

The net effect of all that lobbying has been immensely profitable for the oil industry—$340 million in lobbying "investments" over a three-year period, in exchange for more than $30 billion in tax give-aways over the next ten years, or an overall payoff ratio of almost ten to one. Few but the luckiest of the Texas wildcatters ever managed to get rates of "return" that were that high, and most of the wildcatters died broke. By contrast, when Tony Hayward was relieved of his duties as BP's CEO in July of 2010, even though he was basically sent to Siberia—technically, being offered a post at the company's TNK-BP joint venture in Russia—he was able to take along a "golden parachute" from BP, amounting to more than $1.5 million in salary, plus a pension worth over $17 million.

What the United States gets in return, unfortunately, is not so impressive. Not only do we inhabit a world where our remaining reserves of precious petroleum are disappearing fast, but we receive a lower rate of payment from oil companies for those reserves than almost any other county in the world. Finally, as if that is not enough, we have allowed oil companies to exert such powerful influences over the writing of tax laws that we encourage them to drain America's oil even faster—effectively discouraging investments in other lines of business at the same time. It's quite a deal.

9 Cleaning Up

Déjà Vu All Over Again

Some forty years ago, after the Santa Barbara oil spill, a friend of ours wrote about the striking disjuncture between the technology available for oil drilling—already sophisticated and expensive at that time—versus the distinctly low-tech options that were available for cleaning up the spill. The drilling and production were being carried out with some of the most precise equipment ever invented, while the so-called clean-up was being done with straw, rakes, shovels and garbage cans. During most of the time when we were writing this book, efforts were still underway to stop and "clean up" the blowout from the BP disaster, and watching the ineffectiveness of the so-called "clean-up" techniques made his points even more striking, forty years later.

Various efforts were made to stop or control the spewing oil, but from the start, industry experts predicted that the only true hope for finally stopping the ongoing tragedy would come from drilling a new, "relief" well. Making use of today's extraction technology, that relief well was designed from the start to be drilled through a mile of water, then through another two and a half miles of sediment and rock—all then to hit a target that is only inches in diameter. In percentage terms, this kind of drilling technology is actually more precise—and in our view, more sophisticated—than doing brain surgery on a single cell, blindfolded.

The available *clean-up* technology today, by contrast, is little more advanced than the straw bales and shovels of forty years ago.[1]

Two decades after the Santa Barbara spill, but three weeks after the *Exxon Valdez* spill, one of the authors of this book stood with other members of a National Research Council committee, watching an almost surreal "clean-up" activity that was being conducted by Exxon on the beach of one of the smaller islands in Prince William Sound. A couple of dozen men in brightly colored jump suits were scurrying about the beach at low tide. Four points about their activities are important to understand. First, the "beaches" in Prince William Sound consist mainly of gravel and flat, gray stones, most of which are about the size of a child's hand. Second, this activity was necessarily confined to low tide, because the tidal range in Prince William Sound is about ten feet, and at high tide the beach would be covered by water. Third, while Prince William Sound is more protected than the Gulf of Alaska, to which it connects, it can still be a challenging body of water, with waves that are not to be taken lightly. This means that as the tides rose and fell, the beaches would not only be submerged, but be subjected to wave action that could be quite intense. Fourth and finally, because on this particular day there was a gentle but steady wind, these men needed to work on the lee side of the island to avoid the waves.

The actors and their props were positioned as follows. Two garish orange floating booms were stretched in a V, from a section of the beach to the sides of a skimmer boat. The boat had a submerged, ramp-like front with a conveyer belt arrangement that was designed to lift floating oil and deposit it in a storage area. Within that V, a pump was sending seawater up on the beach—the same beach that would soon be submerged by the same sea water—and allowed to run down to the water's surface. This produced a very light sheen—not visible oil or globs of tar, but a microscopically thin sheen—which the skimmer was attempting to lift off the water. Tellingly, there was no storage barge, or facility, to hold the contents from the skimmer after it was full. Other members of this strange drama were actually wiping down rocks above the tidal line with paper towels. The futility of the whole operation prompted one member of the committee to comment that what we were watching was an entirely political activity, not an environmental one.

Twenty-one years later, the same author stood on the beach at Grand Isle, Louisiana, about a month after the BP *Deepwater Horizon* blowout, watching as a couple dozen men—this time in white jump suits—scraped at the beach with square-point shovels, collecting tar balls and depositing the balls in plastic bags. They were being followed by an enormous front-end-loader. After collecting their meager prizes, they would drop the bags in the bucket of the loader. Tellingly, there was still nothing else into which the loader could dump the accumulated contents once it was full. Again, this drama was played out at low tide, because at high tide, this area of the beach would be covered by water—and the high tide would bring more tar balls. The beach at Grand Isle is sand, instead of gravel and stones, and the tides seldom exceed a range of three feet, but the futility was the same as two decades ago.

Several days later, the same author held a conversation with an engineer at his university. The response to the BP blowout came up, and the engineer mentioned that he was involved in oil spill remediation research following the *Exxon Valdez* spill. After a while, the interest, and research dollars, dried up, so he went on to other things, until he went back to check on the recent literature in the aftermath of the BP blowout. As he recounted it, "The good news is, I'm still an expert. The bad news is, after eight or nine years, I'm still an expert."

The even worse news is that this lack of progress in cleaning up an oil spill is taking place in a world where we still really don't know how to clean up the oil that has been spilled into water bodies of any size. While the old adage is that oil and water don't mix, we have yet to invent the techniques that can separate them, particularly after they are combined in significant quantities in large bodies of water. Oil spills are not new phenomena, but in no case—from the *Torrey Canyon* to the *Amoco Caldiz*, from *Ixtoc* to the *Odyssey*, or from the *Exxon Valdez* to BP and the *Deepwater Horizon*—has anyone ever been able to get more than about 5–10 percent of the oil back in the boat. That, moreover, is a generous estimate.

Skimmers work well with thick oil in swimming pools, and booms can cordon off oil spills in small ponds, *if* there is no wind. In Prince William Sound and the Gulf of Mexico, though, they have proved to be essentially useless. In addition, mixing dispersants with the oil distributes the oil across a range of water depths, defeating the whole theory behind skimmers and booms, which will only work on floating oil. Finally, there is not even a theoretical possibility that these primitive technologies would work within the Louisiana coastal marshes. While a small percentage of the rocks on the beaches in Prince William Sound could at least be wiped off by paper towels, consider cleaning individual blades of marsh grass over dozens or even hundreds of square miles of marsh, and the futility of the situation, or indeed its stupidity, makes itself apparent.

Even after the "kill" operations finally managed to shut down the gusher of goo from the bottom of the sea, the environmental and social damage continued to mount. A review of daily wildlife rescue reports by the New Orleans *Times-Picayune*, several weeks after the initial capping of the well, found nearly a doubling in the number oiled birds being "rescued" each day—except that the fraction of the birds being recovered *alive* actually went down, from an average of 56 percent before the capping to just 41 percent afterwards. The figures for sea turtles were even worse, with more turtles having been recovered in the few weeks after the capping than in the several months before. By August 2010, government officials would declare that the vast majority of the oil from the spill had been burned, skimmed, recovered, or "dispersed," but independent scientists would disagree strongly, concluding that as much as 80 percent of the oil remained in the Gulf. One major study from the Woods Hole Oceanographic Institute, based on data from June, would report an underwater plume from the BP spill that was 22 miles long, more than a mile wide, and as deep as a 60-story office building is tall. As noted earlier, however, debates over the actual severity of the spill can be expected to go on for years—perhaps exacerbated, in this case, by the extensive use of chemical dispersants, which may have helped to keep the oil from rising to the surface, but which also may or may not have introduced other dangers to the marine ecosystem.[2]

What is not in dispute is that BP's Oil Spill Response Plan was spectacularly at odds with reality. BP claimed that it could "handle" not just a worst case spill—which BP estimated on the basis of a nearby lease to have a potential volume of 250,000 barrels per day, meaning 10.5 million gallons of oil, or about the quantity of a new *Exxon Valdez* spill each day—but that the company would be able to activate nearly twice that much containment and clean-up capacity, or about 491,721 barrels per day. In fact, as Americans could see on their television sets, day after day and month after month, BP was unable to come close to "handling" the actual spill of about 60,000 barrels per day.

What is even more worrisome, though, is that the federal agency that was charged with protecting America's natural resources from such oil spills chose to approve BP's plan. A number of observers have suggested that the very idea of allowing such a statement to be incorporated into a formal planning document, which someone else would then approve, amounts to something between incompetence and criminality. The point comes too late to help the Gulf coast this time around, but it will clearly still be relevant for the next spill: The name of the game needs to be prevention, not a mistaken belief that we can actually "clean up" such a mess.

Moving Out of the Past

It is high time that we started to expect our political leaders to be more realistic. At a minimum, we need to acknowledge that, if oil is spilled into the marine environment, the overwhelming majority of it will stay there until natural processes remove it, however long that takes. What this means is that existing response documents are philosophically backward, or in ordinary language, useless. The so-called contingency plans that exist at present amount to what Clarke has called "fantasy documents," and the reclamation process, which will proceed with the same painful trial-and-error approach that characterized the first three months of BP's efforts to stop the oil eruption, will probably need to go on for a very long time.[3]

More broadly, there are at least three key principles that need to become integral and unquestioned parts of our risk-management systems in the future. The first is as old as the principle that it is never wise to put all of one's eggs in one basket. If the last thing we need to be doing is writing fantasy documents—then stumbling around as we learn, in a time of crisis, that their claims are either impossible or ludicrous—the second-to-last thing we need is to be relying on a single line of defense, particularly if that defense is provided by so-called blowout preventers. Since we know that blowout preventers can't actually be relied upon, we should avoid any situation where we need to rely on them. In many cases, blowout preventers have worked to "shut in" a well, but, as indicated earlier, this very fact may have contributed to an atrophy of vigilance, or in simpler terms, an over-reliance on the hope that this last mechanical line of defense would actually do the job. While we are not familiar with any published failure-rate data, engineers we know who are experienced in deepwater drilling do not trust blowout preventers—and the blowout of BP's Macondo well indicates the wisdom of their judgments.

It is also possible to offer suggestions for improvement that would not have been that complicated or difficult to have in place. One illustration has to do with the last line of defense *in* that last line of defense, namely the device in the blowout preventer called a *shear ram*. While there are other hydraulically operated rams that close around the drill string and seal the well, the shear ram is intended to work with brute force. It has two blades that interlock, theoretically cutting through the drill string, and anything else in the way, to seal the well. One key problem with that plan, though, is that if the shear ram happens to hit one of the joints that connect the lengths of drill string to one another, it won't be able to cut all the way through. This is not something that would strike anyone in the oil industry as a shocking new insight, and in fact, one possible solution to this problem is so simple that it is already in place on the majority of rigs operated by Transocean—but not on the *Deepwater Horizon*—namely to include two shear rams in each blowout preventer, separated by a distance that is greater than the length of a joint in the drill string.

Second, we also need to improve the odds. Since there may well be future cases where other lines of defense will fail, we must also take other steps to improve the performance of each step in the process. Useful illustrations can be drawn from a number of other industries, such as the airline industry. That industry, too, sometimes falls short of perfection, but on a jet engine assembly line, each part of the emerging engine is traced in a document from its original starting point (metal pour, plastic batch mix, etc.), through each process (shaping, milling, molding), to its final placement in the engine. One copy of the resulting, exhaustive manual accompanies the engine, and one copy is filed with the manufacturer. If a part in an engine later fails, all of the engines with a part manufactured by the same process can be identified and inspected.

At a larger level, moreover, if the airline industry is far safer today than it was twenty or forty years ago—and it is—part of the reason is that, for the most part, airline maintenance scheduling is rigid and unyielding, check-off procedures are extensive and required, and there is a great deal of emphasis on doing things properly, every single time. If an airline mechanic says "something isn't right here, and we need to figure it out before we proceed," it is highly unlikely that anyone in the chain of command would override that opinion—perhaps in part because one of the obvious people in the chain of command is the airline pilot, whose own life might well be on the line if the mechanic's views were to be ignored. For similar reasons, engineers are often expected to be some of the first to test the results of their efforts. The uncle of one of the authors of this book, for example, was riding on top of the first undersea storage funnel to be towed into place and then sunk to the bottom of a shallow sea, off the coast of Dubai. His team's calculations—just how far the structure would lean before "burping" out the air that was under the funnel, how far and high the water would spray, how strong the structure needed to be—proved to be right on-target. It would not be possible to have an engineer riding a blowout preventers down to the bottom of the ocean, but that and all of the other machinery associated with an offshore rig need to be given the same kind of maintenance and scrutiny.[4]

Still, the third problem with a focus on blowout preventers, or any physical part of an engineered system, is that this is not where the problems usually lie. An engineer with long-time experience in the offshore industry put it this way:

> My initial assessment of the root cause of the MC252 incident is failure by BP management to manage a series of technical and operational changes. Failure to conduct a proper hazard analysis on the first change was a secondary cause. More information might change this initial assessment but that information may never come, due to the extreme difficulty of gathering onsite data. Killing of the well may obliterate any real information about the technical cause of the blowout.
>
> MC252 is not an operator or mechanic root cause but a corporate management root cause, pure and simple.[5]

Another engineer, equally experienced, reminded us that the causes are not always top-down in nature:

> There has been much discussion of the immediate technical causes of the 2010 BP oil leaks, but the underlying or root cause was a change in the culture of the company as a whole, particularly in the U.S. Back in the 1960s and 1970s, BP had a good safety record and safety culture. Why has it changed?
>
> Every manager in the chemical and oil industries (and no doubt many others) knows that many operators, maintenance workers, foremen and shift managers ... will start taking short cuts or stop following instructions if they can get away with it. Their attitude is a macho one, that their job is to make product, not fill in forms such as permits-to-work. Professional staff, at all levels, should check details frequently or the procedures will corrode faster than the steelwork and soon vanish without trace. It is not enough for the professional staff to take a helicopter view.

On many, perhaps most, offshore rigs and platforms, the senior person in charge is usually someone who has risen from the ranks and would have been a foreman or shift manager if he had stayed onshore. While professional staff visit from time to time, there is no day-to-day checking. When there is a problem, the attitude of the senior person on the rig or platform is often a macho one: "Forget (or a shorter word) the rules. Let's get stuck in and finish the job."

It is easy to point the finger at the management and assume that a culture of cutting corners started at the top, and was motivated by money. It's worth remembering that the same culture can also originate at the bottom, driven by the desire to get the job done. The task of management is to know this and make sure it is done properly.[6]

In an article that specifically examines the occurrence and prevention of blowouts—which are still the single greatest cause of major accidents on offshore platforms—engineering professor Bob Bea identifies organizational and personnel issues as the major culprits:

The majority of the organizational factors are associated with conflicting incentives that are provided to the drilling and workover personnel. These conflicting incentives most often are those of production and safety. These conflicting incentives are not unique to drilling and workover operations in this industry or even to the oil and gas industry.[7]

In a similar vein, the organizations expert Charles Perrow had this to say in comparing airline and marine systems:

As long as a captain meets the production level expected, no action is taken even if it is known that he takes large risks to do so. If the captain falls below this production level, pressure is increased. If the result is an accident, the captain is blamed, and penalized through fines or dismissal.[8]

One of the overriding factors seems to be transparency. As Perrow points out, the airline system operates with direct contact with the public, including that portion of the public that includes airline executives and national political figures, essentially all of whom are frequent flyers. Airlines keep meticulous data on weather and other delaying factors, and the planes have "black boxes" that record various actions of the pilots and configuration of the plane. Traffic and route decisions are made by ground controllers and are also recorded. This produces a system where management doesn't have to trust personnel accounts of why a plane was delayed or why an action was taken. They can know.

In contrast, the captain of a ship operates far from public surveillance or regulatory oversight, usually doing so in a context where the pressures to meet schedules will conspire to produce what Perrow calls an "error-inducing system." Ironically, in error-inducing systems, the authoritarian structure of the operating system—virtually all authority lies with the captain—is sometimes seen as an advantage, at least when emergencies arise, because one decision-maker can react quickly and decisively. With the *Deepwater Horizon*, however, even this decision structure was apparently not present. Emerging testimony recalls arguments between BP and *Deepwater Horizon* management on the appropriate course of action.[9]

In another essay, in which we suspect his words were slightly tongue-in-cheek, Bea also noted,

> Two things are the bane of engineers: uncertainties and people. Uncertainties devil the engineer because his designs must be deterministic; certain. But, the world is uncertain and the engineer constantly struggles with how to cope with the uncertainties. People devil the engineer because fundamentally they are not predictable, and often not controllable. They do not fit easily into engineering equations and analytical models. In addition, most engineers "want to believe that the planet is not inhabited." The history of failures of engineered systems clearly show that it is these two things that are at the heart of failures of most engineered systems.
>
> ...To many engineers, the human and organizational factor part of the challenge of designing high quality and reliability systems is not an engineering problem; frequently, this is believed to be a management problem. Often, the discrimination has been posed as technical and

non-technical. The case histories of recent major failures clearly indicate that engineers have a critical role to play if the splendid history of successes and achievements is to be maintained or improved. Through integration of technologies from the physical and social sciences, engineers can learn better how to reach such a goal. The challenge is to apply wisely what is known. To continue to ignore the human and organizational issues as an explicit part of engineering is to continue to experience things that engineers do not want to happen and whose occurrence can be reduced.[10]

In a more serious comment, the Deepwater Horizon Study Group—pulled together in part by the same Professor Bea—offered a far more concise assessment: "This disaster was preventable had existing progressive guidelines and practices been followed. This catastrophic failure appears to have resulted from multiple violations of the laws of public resource development, and its proper regulatory oversight."[11]

Making Changes

As this discussion may make clear, we do not see the most immediate need for reform as lying in the technology of offshore drilling, even though advances in the technology are easy enough to identify, and they do need to be implemented. The bigger problem, instead, involves humans, not just hardware. While operations like the BP Macondo well appear to be at the very limits of our technological abilities—or even a little beyond, into the realm of the technological Peter Principle—it is already evident that the BP blowout was an organizational as much as a technical failure. Here, the ways to make improvement are not as simple as adding a second shear ram to future blowout preventers. Still, at least three approaches have been suggested as having the potential to reduce the probability of another blowout like Macondo—exclusion, regulation, and refocusing.

Exclusion

The simplest of the three is the idea of exclusion—bad actors should not allowed to play. Given BP's record of safety violations, as discussed above, this approach would argue that the company should be excluded from participation in an activity that is on the edge of our technological ability, such as deepwater drilling. Under such proposals, BP could be required to hand over the direct control of drilling operations to minor partners (such as Anadarko with the MC252) or to the drilling company (Transocean in the case of the MC252), or to relinquish its leases, which could be bought back and offered for bids in future auctions. At a minimum, BP could be denied certification for future lease auctions, at least until drastic realignment of their drilling program and a supervisory arraignment would be negotiated. Possible variations of this approach might involve some sort of probation period, as well as a requirement that someone other than a BP employee would have to be able to veto risky behavior.

Such proposals have an obvious appeal, but in practice, they tend to have equally obvious limitations. While it is easy enough to see the attractiveness of such a proposal in the aftermath of a disaster such as the BP blowout, "bad actors" are always easier to identify during the week after a disaster than the week before—and the track record of federal regulators in identifying or taking action against "bad actors" is scarcely the sort of thing that inspires confidence. Recall that BP had avoided being black-listed in Alaska by agreeing to a kind of probation—not that it seemed to make much difference in actual organizational behavior. Recall also that the company been named as a finalist in the Department of the Interior's 2009 competition for the MMS Safety Award for Excellence (SAFE), and it would have a strong contender for top honors in 2010 except for one small problem—the fact that, just two weeks before the awards ceremony, its operations on the *Deepwater Horizon* blew up and sank, creating the largest peacetime oil spill in history.

Regulation

A second approach, accordingly, is to consider new approaches to regulation and oversight. The most common proposals call for moving regulatory responsibilities from one agency to another, but a more creative alternative, developed after the *Exxon Valdez* spill, involves the requirement for the industry to provide funding support for independent oversight from ordinary citizens. Under the Oil Pollution Act of 1990—the law that limited oil companies' liability for economic damage from an oil spill to a laughably tiny $75 million—two Regional Citizens' Advisory Councils have been established in Alaska to monitor the safety of tanker and terminal operations, to promote improved environmental and public health outcomes, and more broadly, to counter the kind of creeping complacency in industry and government agencies that contributed both to the *Exxon Valdez* spill and the BP *Deepwater Horizon* blowout.

The councils include representatives from key stakeholder groups that are rarely at the table when oil industry representatives meet with government officials, such as commercial fishing interests, tourism, indigenous groups, and local communities. At least equally important, though, is the fact that the councils have industry-supported budgets to commission independent scientific research and to hire professional staff to carry out much of the day-to-day work. It is not clear whether a similar model would have been effective historically along the oil-dependent stretches of the Gulf of Mexico, given the strong social multiplier effects of the oil and gas industry, but in light of the seriousness of the damage from the BP blowout, it does appear that such a model would have a good deal to offer in the future. In Alaska, levels of funding that are trivial by oil industry standards but high by standards of citizen organizations, adding up to several million dollars per year, have contributed to a number of improvements in safety and environmental outcomes, ranging from requirements for escort tugs and increased use of double-hulled tankers, to reductions in dangerous air pollution from the oil terminal, and efforts to help other communities to learn how to cope with the fact that an oil spill can have devastating social as well as environmental consequences.[12]

The more common approach in the past, however, has been to focus on direct federal regulation, and in this arena as well, it is clear that there is considerable room for improvement. Most workplace safety, for example, is under the jurisdiction of the Department of Labor's Occupational Safety and Health Administration (OSHA). Because drilling rigs are currently classified as "vessels," they fall under the Coast Guard, which coordinates its inspection program with the Minerals Management Service (MMS) through a memorandum of understanding. The net effect is that OSHA currently has no regulatory or enforcement authority over mobile oil drilling rigs or production platforms on the Outer Continental Shelf, three miles or more offshore, which is where the *Deepwater Horizon* was located.[13]

Starting after the July 4th weekend in 2010, clicking on the bookmark for the Minerals Management Service Web site brought this announcement:

The New Bureau of Ocean Energy Management, Regulation, and Enforcement (BOEMRE) Web Site is Under Construction

Secretarial Order 3302, issued June 18, 2010, renames the Minerals Management Service to the **Bureau of Ocean Energy Management, Regulation, and Enforcement (BOEMRE)**. The name change is effective immediately.

This is certainly a large symbolic gesture, but one wonders if it will bring about any real changes in safety. Personnel from MMS have been inspecting offshore operations for many years, and they have long-time associations with personnel on the facilities they are inspecting. If the same individuals are inspecting the same operations, will it matter if the letterhead on the form reads BOEMRE instead of MMS? On the other hand, if the Coast Guard can operate under a memorandum of understanding with MMS, or BOEMRE, then they could certainly operate under similar understandings with OSHA.

There are at least three potential advantages in the idea of asking OSHA to take over duties for offshore inspections. First, rather than having conflicting goals of encouraging production and safety, OSHA has no official agenda except for workplace safety. Second, OSHA has extensive experience with regulating dangerous industrial processes, including those in oil refineries. Third, OSHA has codified federal regulations in place, called Process Safety Management for Highly Hazardous Chemicals (29 C.F.R. 1910.119), intended to prevent or minimize the consequences of a catastrophic release of toxic, reactive, flammable, or explosive hazardous chemicals.

At the same time, however, the idea of reassigning responsibilities, rather than simply renaming an agency, is also one that has potential problems, making it unwise to believe that such a rearrangement of responsibilities would somehow solve all problems in regulating industries that can benefit financially, at least in the short run, from cutting corners. Any agency can be subject to "capture" by the industries being regulated. It is also certainly possible that a future president would choose to appoint top-level personnel to OSHA who would focus mainly on making life easier for regulated industries. Particularly in the case of industries that measure their annual lobbying budgets in tens of millions of dollars, it is not difficult to see the potential for Congress to pass new laws or limits on agency expenditures to avoid "needless" regulatory actions that oil-friendly members of Congress might find objectionable.

In the wake of BP's Texas City refinery explosion, moreover, OSHA *was* the agency in control. Its leaders even initiated what they called their National Emphasis Program (NEP), stepping up inspections of refineries. In testimony before the Senate Committee on Health, Education, Labor and Pensions, unfortunately, the deputy assistant secretary for OSHA, Jordan Barab, gave a less-than-glowing report of the actual consequences for safety:

> I am sorry to report that the results of this NEP are deeply troubling. Not only are we finding a significant lack of compliance during our inspections, but time and again, our inspectors are finding the same violations in multiple refineries, including those with common ownership, and sometimes even in different units in the same refinery. This is a clear indication that essential safety lessons are not being communicated within the industry, and often not even within a single corporation or facility. The old adage that those who do not learn from the past are doomed to repeat it is as true in the refinery industry as it is elsewhere. So we are particularly disturbed to find even refineries that have already suffered serious incidents or received major OSHA citations making the same mistakes again.[14]

As readers with healthy levels of skepticism may recognize, such a pattern of problems may go deeper than "communication." The simple fact of the matter is that corporations seek to maximize profit, and—at least before or until a disaster strikes—safety measures can look to those corporations as though they are "needless" costs. Making matters worse, it is of course possible to send the CEO to Siberia, literally or figuratively, but in the complex, interlocking relationships found in the corporate world, it can be difficult or even impossible for outsiders to determine responsibility—and hence to assign individual blame—for a particular action, or lack of action. That is particularly so in light of the fact that a cost-cutting action (or inaction) that violates safety rules is a gamble, which may or may not be caught—and even if it is caught, the fine is almost always less that the profit that was obtained by cutting corners. Unless these conditions change, Mr. Barab or his successors in OSHA are likely to be making the same kinds of remorseful statements in future Congressional hearings.

If neither exclusion nor regulation provides all that much in the way of reassurance, that leaves us with refocusing—meaning the need to recognize that we may need to look not just at our government bureaucracies, but at the ways in which we have come to live our lives.

10 Today and Tomorrow

Contrary to the superficial impression that expanded offshore oil drilling would be "good for the economy," the reality is that U.S. energy policies over the past quarter-century have conferred most of their benefits to a handful of the world's largest oil companies, doing so while offering little if any visible advantage for the larger economy, and clearly creating losses for the federal treasury—continuing to do so during decades of record federal budget deficits. Rather than "helping the economy," federal policies appear to have emphasized the transfer of valuable, resource-rich undersea lands from the general public to a handful of the richest corporations in the history of money. Unfortunately, the existing management model, with its low royalty rates and practical limitations on participation by smaller companies, has done significantly less well in bringing benefits to the legal owners of offshore resources—the citizens of the United States.

Today, there are some 3,500 offshore production facilities in the federal waters of the central and western Gulf of Mexico, about three quarters of them off the shores of Louisiana. They are accompanied by over 25,000 miles of buried pipelines, connecting the platforms to each other and to the shore. The evolutions of this massive industrial network over the six decades since Kerr-McGee's first successful well in 1947 has radically transformed the social as well as the biophysical environment of coastal Louisiana, and to a lesser extent, that of Texas, as well.

At the time when Interior Secretary Watt was pushing for even more aggressive efforts to offer industry the oil reserves along the nation's coastline, even the state of Louisiana objected. At the direction of the state's then-governor, Chip Groat, director of the Louisiana Geological Survey, sent a letter to John L. Rankin, U.S. Department of the Interior, Bureau of Land Management, noting:

> The State of Louisiana is concerned about the potential impacts of proposed changes in OCS leasing procedures that are designed to open the entire OCS area to exploration and development. Under the proposed leasing schedule it is likely that the industry would concentrate its initial efforts in areas of proven production such as the central Gulf. The resulting intensive development would create severe economic and environmental impacts in coastal Louisiana. This would also lead to a major increase in the rate of depletion of our most productive OCS area which is not in the best long term interest of the United States.[1]

History had by then provided over a century's worth of evidence that the concerns in the letter were on-target. In Pithole, at Spindletop, and in hundreds of other boom-and-bust regions, what had taken place was indeed a pattern of "severe economic and environmental impacts," during a brief if sometimes exhilarating period of booming development, coupled with a high "rate of depletion," leading all too soon to a painful bust.

The high-level officials of the Department of the Interior that we knew at the time, however, disagreed quite strongly. They claimed that area-wide leasing would lead to the kind of expansion in offshore oil production that had been promoted by every president since Richard Nixon, that it would lead to a substantial increase in the revenues that would flow into the

U.S. Treasury, and that oil reserves off the shores of Louisiana were so vast that there was no real possibility of depletion, let alone a bust.

More than a decade before Secretary Watt took over, at the time of the Santa Barbara oil spill in 1969, the effort to extract oil under provisions of the Outer Continental Shelf Lands Act had already been underway for some fifteen years. Annual crude oil production on the Outer Continental Shelf had risen almost a hundred-fold from its 1954 level of 3.3 million barrels, climbing to 312.9 million barrels, or almost a tenth of total U.S. production. That, however, was to be nearly the high-water mark, and almost all of the OCS oil was still coming from the Gulf of Mexico. Although Secretary Watt had hoped to increase Outer Continental Shelf production from "frontier" regions—those outside of the Gulf of Mexico—his efforts effectively inspired Congress to impose moratoria, stopping the rapid growth that federal officials had hoped to see from those other locations. Gulf production, meanwhile, actually peaked in 1971, and it had already been in decline for the better part of a decade by the time when the state of Louisiana expressed its concerns about "a major increase in the rate of depletion of our most productive OCS area," noting as well that draining this precious resource even faster was "not in the best long term interest of the United States."[2]

As local officials feared, the boom was indeed followed by a bust—one that lasted for a decade or more. Louisiana, however, has one geographic advantage not enjoyed by Pithole, or Spindletop, or most of the other energy regions that have gone through boom-and-bust experiences in the past. As exploration moves on to other fields, further offshore, oil companies have good reason to stay put, since there are no other patches of reasonably dry ground that are any closer to the deepwater fields than are the existing facilities along the Gulf coast. Unlike most earlier busts, accordingly, the long Louisiana bust of the 1980s and 1990s eventually came to an end, and growing numbers of people once again started to express optimism about their economic prospects. This time, on the other hand, the region's residents warned repeatedly, "just don't call it a boom."

By the time the drilling started on BP's *Deepwater Horizon* well, it would have been fair to say that the oil industry in Louisiana was mostly "back." Unfortunately, it was also showing signs of "turning its back." Many Louisiana oil company offices have been closed down, with the operational centers of the industry moving toward Houston and other cities. On the North Slope of Alaska, oil extraction is continuing, but there, too, the coming of the end is a matter of when, not if.

After BP's Macondo blowout, business observers speculated that BP—still one of the key investors in North Slope oil—might need to sell its share of Alaska oil fields to help pay off the costs of its massive spill and cleanup in the Gulf. As noted in some of the reports, even the massive Prudhoe Bay oil field has long been on a downward trend. The field produced 2.1 million barrels a day in 1986, but it was down to a production capacity of 400,000 barrels per day in 2010—and in the month when BP was engaged in talks to sell its interests in the field in mid-2010, the actual level production had dropped to just under 235,000 barrels per day. That means that the volume of oil being spewed into the Gulf from BP's one "nightmare well" at Macondo amounted to about a fourth as much as the *total* production from the Alaska north slope at that same time.[3]

Today, official estimates hold that planet Earth still contains about a trillion barrels of proven reserves. In rough numbers, and at today's rates of use, that amount would be about enough for 30 more years of consumption.

Two points need to be kept in mind when thinking about any such estimates. On one hand, the "30 years" estimate is not much different than the estimates that were made 30 years earlier; some of which also called for proven reserves to last for only about three decades. Given that the oil is still flowing today, it is obvious that something kept those earlier predictions from proving accurate. One key "something" is the fact that, in interim, the oil industry has managed to do what its leading spokespersons had predicted a third of a century earlier, which is to find "new" reserves. The other key "something" we need to remember, though, is that these are not really "new" reserves—they are ancient. What the oil industry has actually done is to find a higher fraction of the ultimately finite deposits that were bequeathed to us by the era of the dinosaurs. We are finding more of the oil—but the dinosaurs and the environments they lived in are not making any more of it. As even school children learn, after all, the dinosaurs have long since become extinct.

It was in a T-shirt shop in Valdez, Alaska, many years ago, where we first encountered a saying that we have since seen many times in Louisiana: "Please, Lord, send us another oil boom. We promise not to piss it away this time." For the United States as a whole, of course, the "oil boom" has lasted far longer, but that fact is not likely to confer any immunity to what can happen if we continue to fritter away a valuable but finite resource. At some point, the day of reckoning will come.

Today, the world's "new" oil supplies are being found in some of the least convenient locations known to humanity—some because of geography and climate, others for reasons of geopolitics. Pithole had a small reservoir, but one that came to an end. Louisiana and Alaska have larger reservoirs, but both of those are well on their way to reaching their ends as well. The oil-bearing province known as planet Earth has the largest oil reserves of all, but even those reserves are already showing signs of following the trajectories of the others.

For roughly a century—from the time when the first inventive Americans thought up the idea of drilling into the ground in search for rock oil in the 1850s, until roughly a century later, in the 1950s—the United States was the dominant producer of oil on the entire planet. No less profoundly than oil production has shaped the economies and the ways of life along the south edge of Louisiana and the north edge of Alaska, that heady history of oil dominance has also shaped the ways of life that Americans have come to see as "normal." During the even more heady times that followed the end of World War II, American political leaders unfortunately took the nation further along the pathways that had been established over the previous decades. Having learned to see normalcy as a fast trip on a river of oil, the Americans who were born after the end of World War II were actually coming into a world where nothing could be less normal. Perversely, moreover, the more we act as if the world's oil supplies are endless, the more rapidly we are guaranteed to bring those supplies to an end.

At present, the world still contains additional territories to be discovered, and there are good reasons to expect the industry to identify new options for squeezing more oil out of the formations that have already been found. Still other possibilities will be identified by especially innovative thinkers of tomorrow. It is thus certainly possible that the world's oil industry will continue to experience the same kinds of successes over the next several decades that we have come to expect from the past—finding enough "new" supplies to meet the demands of present-day Americans, plus the new Americans who will want even more of this precious substance in the future—all while also meeting the rapidly growing demand for oil from the rest of the world. Another possibility, unfortunately, is that the oil industry's ability to stay ahead of growing demand—while impressive—will soon be running up against not just the limitations of human technology and ingenuity, but those of geology. If we use up the fuels at a much higher rate than new fossils are being created—which is in fact what we have been doing for the last century and a half, at a rate that is nothing short of spectacular—we are creating for the entire globe a process that, like the extraction of oil from any given region, can only go on for so long.

By about the time Katrina was striking the Gulf coast in 2005, the China National Offshore Oil Corporation (CNOOC), owned in significant part by the Chinese government, was bringing a different kind of shock to the U.S. oil industry. In recent decades, the oil industry has seemed to be headed back toward the kind of concentration that led to the Sherman Antitrust Act, as giant corporations absorbed one another in the process of becoming still more gigantic. Exxon merged with Mobil, Chevron merged with Texaco, and then the combined ChevronTexaco corporation worked out an agreement to purchase another oil giant, Unocal—the former Union Oil Company of California, which at the time was the ninth-biggest oil company in the United States. An unexpected CNOOC bid, however, suggested that the future might not be just a replay of the past. Chevron had agreed to buy Unocal for $16.65 billion, and CNOCC offered $18.5 billion. Since 2005, China's economic power has continued to grow—the country recently surpassed Japan as the world's second largest economy—raising the clear possibility that the day will soon come when the U.S. can no longer afford to consume the oil it wants, whether we are talking about the economic, the environmental, or the geological forms of affordability.

The world's oil supplies, while vast, are nevertheless finite. Human abilities to find the deposits that do exist, while impressive, will ultimately run up against the fact that there is only so much of the stuff to go around. Even if the residents of a future Saudi Arabia, or Kazakhstan, are willing to sell their oil to the future residents of North America and Europe, rather than to the future residents of China and other nations—possibly even down to the very last barrel—there will ultimately come the time when the planet's reserves will come close to that very last barrel.

We have already extended the search for oil off both edges of the continent and into some of the most hostile political territory on the face of the globe. There can be legitimate debate about how much further that process can be extended. There can be little debate, however, about the fact that the it cannot go on forever—or about the fact that, by now, it would be more rational to start preparing more rationally for that future, rather than simply to pine for the past.

Facing the Future

In June of 2008, George W. Bush removed the presidential ban on offshore drilling that had been put in place by his father. With election-year politics heating up, politicians around the country started to chant "drill, baby, drill," and Congress allowed its own moratorium to expire just three months later. By the time of the election, agreement appeared to cross party lines, and expanded offshore exploration remained a priority when the Obama administration and the new Congress took office in 2009.

One version of the "drill, baby, drill!" slogan was, "drill here, drill now, bring prices down." Between that time and the BP blowout, prices did decline, dramatically, but that was despite the absence of any noticeable increase in domestic oil drilling. At the same time, even a frenzied pace of drilling in the United

States would be likely to have little to no effect on petroleum prices, for a very simple reason: The United States now produces less than 7 percent of the world's oil, and most U.S. oil deposits have already been exploited. Even if U.S. production could be raised significantly from current levels—a possibility that appears highly unlikely, given that U.S. production has been declining since 1970, despite all of the policies that have been put in place to raise it so far—that would still amount to a tiny share of world oil production. If oil prices do respond to supply and demand, then this tiny increase would do almost nothing to change world oil prices.

Energy use in the United States has gone well past the point where more enthusiastic oil drilling can provide a solution. The United States has a huge petroleum appetite, partly because of policies that were put in place during times of apparent abundance—and that, in retrospect, appear to have been

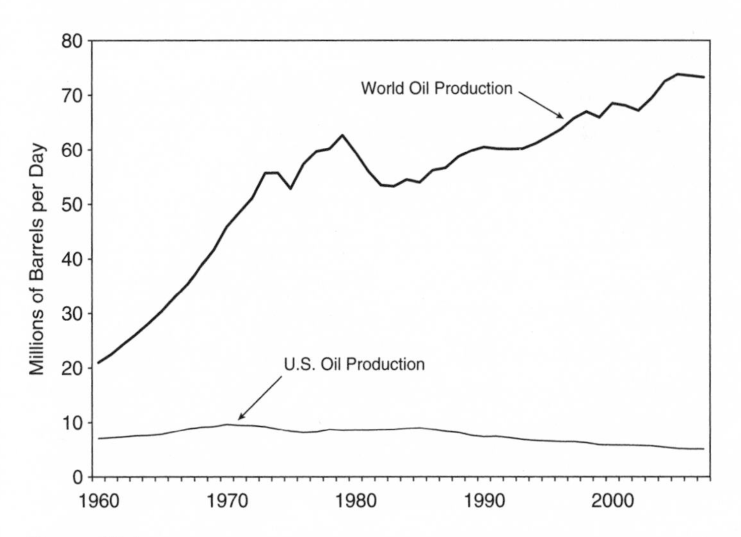

Figure 10.1
U.S. and world oil production

exceptionally unwise. The United States now has less than 2 percent of the world's proven oil reserves. "Energy independence" in petroleum has not been a realistic hope for the United States since before most homes had television sets or, for that matter, perhaps even indoor plumbing, and it will never be possible again. Politicians may continue to make speeches that proclaim the opposite, but in the end, not even the most eloquent politician will be able to negotiate with geology. If all politicians in the United States were lined up end-to-end, they wouldn't be able to produce another single barrel of oil—unless they were first to be covered with sediment for the next several million years or so.

In light of this reality, we have two main options. One, of course, is simply to clench our teeth and to hope for the best. After all, if there is one law in economics, it involves supply and demand. As more and more people "demand" the oil that is becoming increasingly scarce, prices will rise—and economists tell us that, as prices soar, increasing numbers of Americans will decide that their supposed "love affair" with petroleum will become too expensive to continue. The other option, though, is to look to the future more thoughtfully, making the best possible use of the valuable petroleum resources that still remain—both in terms of buying time and in terms of developing new options.

Perhaps this is as good of a place as any to contemplate one last time the dinosaurs that left a legacy, in some of their remains, of the oil we find so valuable today. It is tempting to feel a sense of superiority over the creatures that, while sometimes massive, appear to have had brains about the size of a peanut. Geologists tell us, however, that the era when "dinosaurs ruled the earth" was one that lasted hundreds of millions of years—a span of time that archeologists tell us is much longer than humans, or even human-like ancestors, have been around.

Although the dinosaurs did ultimately die off, the length of their stay on the planet tells us that they were indeed well-suited for the world they inhabited.

That may well have been part of their problem. As a useful simplification, the fine-tuning of specialization seems to be a kind of two-edged sword. On the positive side, the more finely tuned the adaptation, the more efficient an organism can be in exploiting a given ecological niche. On the negative side, a finely tuned adaptation to one environment can prove to be woefully ill-suited if that environment changes. This is one of the key lessons of biological diversity. The problem of the dinosaurs was not that they failed to be efficient enough; it was that they were *too* efficient—too well–suited for the planet they had long "ruled" to be able to survive when that planet suddenly changed.

Next, consider an equally simplified version of what we know of human adaptability. Anthropologists have studied many thousands of cultures, but they have essentially found just one that appears to value "efficiency" over virtually all else—our own. The culture of the industrialized world, with roots in Europe and Asia, and with offspring societies around "the new world" as well as the old, has achieved spectacular successes, resulting in no small part from the fact that the colonizers' societies had relatively high levels of efficiency—along with European diseases and the tremendous energy advantages provided by coal and then oil—allowing them to take over the lands and resources that were formerly the homes of many other cultures that were not so effective at turning available resources into weapons and other tools.

Even so, the new coins that came flowing into the coffers of the colonizers, in earlier centuries, and the industrializers of today, appear to have come complete with two sides. On the one side, "superior" efficiency does seem to have conferred economic advantages on the innovators, at least at the outset, leading to further concentrations of innovation and specialization later on. On the other side of the coin, however, is the simple fact that this one culture, now virtually worldwide, is very recent, and it appears to be unique in human history. Most of the other cultures that survived long enough to be studied by anthropologists were cultures that placed a higher premium on surviving ranges of variation, rather than achieving triumphs in one.

For those who are already adults today, and particularly those who are fortunate enough to live in prosperous, industrialized nations, there may seem to be few reasons for change that an economist would describe as "rational"—meaning essentially "profitable" for the specific individuals in question. For the great-great grandparents of the last generation of dinosaurs, of course, there might have seemed to be equally few reasons to change.

Still, the metaphor of the dinosaurs has at least two limitations for thinking about humans. The first is that the dinosaurs were incapable of planning their futures in advance. The second is that they appear to have been incapable of bringing about their own demise. As Mark Twain once said, "Man is the only animal that blushes. Or needs to." With a bit of adaptation this point can be modified for our purposes: Humans appear to be the only animals that are capable of planning for and shaping the future—but at the same time, given the distinctly human capacity for creating harm as well as benefits, humans are perhaps also the animal that most needs that capacity to look ahead, and to do so with intelligence.

Three forms of intelligence would be particularly well advised:

- First, rather than favoring energy policies that subsidize ever-faster extraction and consumption, we need to favor those policies that can help us to stretch out the supplies that still remain. Energy statistics show, after all, that conservation and improved efficiencies have actually provided more "new" energy then the entire domestic oil industry. The practical American philosopher Benjamin Franklin once noted that a penny saved is a penny earned. The same principle applies, thousands of times over, for a barrel of oil.

• Second, there is an urgent need to use the time that still remains, thanks to our existing but dwindling oil supplies, to develop new ways of providing services that we now obtain from petroleum—heat, light, transportation, chemical feed stocks, and more. At the time of the 1973–1974 oil embargo, one administration observer reportedly mocked the idea of an OPEC embargo, asking ironically what the petroleum-exporting countries would *do* with all of that viscous black liquid—"drink it?" Petroleum may well have been described by all of those snake oil salesmen as having merely miraculous properties of its own, but for most of us, most of the time, oil is useful not so much for what it is, but for what it can provide—and for most of those services, if the proper efforts are made now, we will be able to develop alternatives soon enough to help cushion our grandchildren from the shocks that will otherwise come from the exhaustion of the available supplies. Should this estimate be wrong, and should the oil companies be able to find far more oil than now seems possible, so much the better—we will be able to allow other people and places to worry about their relationships with the Mideast, and with other potentially volatile regions, all while starting to enjoy a form of true energy independence that will simply not be possible so long as we remained so heavily addicted to oil.

• Third and most broadly, we need to end the habit of making the existing reserves seem artificially cheap, tempting ourselves to use up our finite and strategic oil resources even faster, and to end the policy of subsidizing foolish resource decisions more broadly. In fact, given that the region suffering the most from the BP blowout has a good deal of overlap with the one that was struck just a few years earlier by Hurricane Katrina, perhaps it would be worth considering the less obvious kinds of connections that exist between these two types of disasters.

Conclusion: Stormy Weather?

Every summer, as the waters of the tropical Atlantic begin to warm up, what also heats up is the hurricane season. In essence, when pre-existing weather disturbances come together over warm tropical oceans, patterns of rising moisture and wind circulation can create what meteorologists understand to be giant heat engines—and what people of coastal regions understand to be some of the most destructive forces imaginable. The "engine" that drives the whole process is the warmth of the waters, which is why hurricanes are generated only in tropical latitudes, rather than off the coasts of Alaska. As the warm air rises, it spins; as the moisture in that air rises, it condenses—turning into rain and generating still more warmth, in what may be nature's deadliest example of a self-reinforcing spiral.

We think of hurricanes as destructive, of course, because of what happens when they collide with humans and the things we value—threatening a hapless fishing boat at sea, or battering property and costing lives on land. Those of us who have experienced hurricanes know that they can be both awe-inspiring and awful when they hit; partly for this reason, scientists and satellites have spent years studying hurricanes, trying to learn more about how they are shaped and steered, and trying to improve our ability to predict when and where they will make landfall.

Although predictions have improved, the ones that are made several days in advance of landfall are still relatively broad ones. Hurricane X will be predicted to make landfall somewhere between Mobile and New Orleans, for example, somewhere between Sunday evening and Monday noon, with wind speeds that may reach 130 m.p.h. or more. At that point, of course, it would be possible to hold Congressional hearings over whether the more likely time of landfall will be Sunday at 9:00 p.m. or Monday at 9:00 a.m., but not even politicians are generally that silly. Rather than arguing over the exact time or place when the hurricane will hit, sensible people start taking precautions—boarding up windows, evacuating low-lying areas, stocking up on food and water, tying down loose lawn furniture, and more.

Now think for a moment of the gathering storm of the world's oil supply system. It is still possible to argue over whether the entire world supply of oil will lower to life-altering levels in thirty years, or sixty, or for that matter twenty, but save for the ultra-determined few, the people who have looked at the evidence are generally in agreement on the basic facts: What we call oil "production" isn't production, but extraction. Fossil fuels were *produced* by the dinosaurs and other ancient life forms, hundreds of millions of years ago; and we have been using those dinosaur remains so rapidly that the party can't go on much longer. Experts may continue to disagree over whether the energy storm will come crashing up against The American Way of Life by Sunday night or Monday morning, figuratively speaking, but is it really sensible to keep arguing over that range of uncertainty, or would it make more sense to start thinking now about what can be done now, to minimize the damage that will be created when the storm actually does come roaring ashore?

Like all metaphors, of course, this one has its limitations, but perhaps the key limitation is this: The self-reinforcing spiral of Hurricane Katrina was largely out of human control, being created by the warm waters of the Gulf. The self-reinforcing spiral of Hurricane Oil, by contrast, has been created almost entirely by human actions and policy choices. For decades, as the previous chapters have spelled out, political and economic actors in the United States have reacted to almost every problem by trying to speed up the spiral—accelerating the supply of oil through such efforts as area-wide leasing, for example, then trying to boost consumption during the bust that followed a few years later, followed in turn by trying to speed up the rate of supply through so-called "royalty relief," and so on.

For decades, as the ultimate landfall of this energetic perfect storm has stayed beyond the horizon and out of view, this approach has caused few protests from most residents of industrialized nations, who have been happy to keep enjoying the party. This party, however, has been speeding up the coming of the end of the age of oil—and a worsening of the likely impacts of that ending. At each step of the way, at least a few cautious scientists have warned—generally to little avail—that such a speeding up could not after all go on forever, and that, in fact, the faster the spin, the greater would be the destruction when this self-brewed storm will ultimately come crashing against the way of life that we have come to enjoy.

Much of New Orleans is below sea level. It is kept reasonably dry, most of the time, by an elaborate system of pipes and pumps, but Katrina showed us that a direct hit by a hurricane could overwhelm the entire system, putting much of the city, and thousands of its inhabitants, well under water. Most of the time, this possibility is of little concern to the residents, or to the millions of tourists who come each year to enjoy the city's beauty, its parties, and its other charms. When hurricanes come close, on the other hand, all but the hardiest party-goers start to pay a bit closer attention to the question of escape routes. When another hurricane, Lily, came toward New Orleans in 2002—doing so as an extremely severe, category 4 storm, with wind speeds in excess of 140 m.p.h.—the interstate highway out of New Orleans was so clogged with traffic that it took one of our friends 12 hours to drive from there to Baton Rouge, only about 60 miles away.

As Hurricane Lily approached the shores of Louisiana on that October afternoon, though, something important happened. Through sheer chance, just before Lily hit the shoreline, it happened to follow almost the same track as a smaller tropical storm, Isidore, just two weeks earlier. Among its other effects, Isidore had stirred up the waters of the Gulf, mixing the warm surface waters with the cooler waters from below. Without having the warm waters to keep fueling the heat engine, Hurricane Lily's top speeds dropped quickly. By the time Lily crashed into coastal Louisiana, in the early hours of the morning, the storm did so with top wind speeds of 95 m.p.h., not 140 m.p.h.

There may be a lesson there.

Aside from continuing to argue over the exact time when The American Way of Life will be walloped by Hurricane Oil, we still have the chance today to make choices. Roughly half of the homes that Americans will inhabit by the year 2050, for example, have yet to be built, and by choosing to build those houses in ways that minimize commuting—rather than maximizing the continued destruction of former farmland across distant suburbs—we can choose to put new hardware in place that will either worsen or lessen the impact of growing oil scarcity in the future.

To date, however, most of the choices that we and our leaders have made are ones that focus mainly on continuing to enjoy the party in the short run, and to let future generations worry about meeting their needs with whatever remnants of oil may be left. To be fair, a few people who are paying closer attention have started to take the energetic equivalent of boarding up the windows, tying down the furniture, and praying for the best. At present, however, we still have another option—the equivalent of trying to cool the waters, robbing Hurricane Oil of at least the most devastating of the storm potential it might pack when its winds come crashing into our way of life.

For those who may not have read *100 Years of Solitude*, it may be worth reflecting on something else that BP may not have considered when it named its well Macondo. At the end of the book, Macondo disappears—forever. The village gets "wiped out by the wind and exiled from the memory of men." As García Marquez notes, "races condemned to one hundred years of solitude did not have a second opportunity on earth."

It would be overly dramatic to assert, but in the wake of the nonfictional Macondo disaster, it is still worth considering, that if a race has been condemned to one or two hundred years of "solitude" from honestly recognizing the reality of geology, that race—the human race—might also "not have a second opportunity on earth."

We don't go that far. We fully expect that the human race will continue well into the future, whatever the energy choices we make today. The culture of oil consumption that so many of us take for granted today, however, almost certainly will not have a second opportunity on earth. Our preference, accordingly, is to choose rationality and decency over a more sudden and dramatic end to America's hyperdependency on hydrocarbons.

The reasons for making the right choices today are the same ones that have always informed the wisest and most noble choices that humans have made, throughout the history of our species. As beings who have a conscience as well as a consciousness, we hope to leave to our children and our great-great grandchildren the kind of world that our parents tried to leave for us—a world that is better, and one that offers them better reasons for hope, rather than one that we have simply wrung dry.

Now that the storms clouds are pulling closer, and now that the BP blowout has demonstrated in vivid fashion what can happen when we take shortcuts today without worrying about the potential consequences for tomorrow, perhaps the time has come to stop taking the approaches that politicians and oil companies have long favored—continuing to heat things up and hoping for the best. Instead, the choice that would make the most sense would be to start trying to cool the waters and slow the spiral, doing so before getting wiped out by the winds of Macondo.

As we said at the outset, it's about time.

Notes

Prologue

1. On the earlier well, see Coy and Reed 2010. On the safety celebration, see Breed and McGill 2010. See also *New Orleans Times-Picayune* 2010a.

2. Quotations are from Breed and McGill 2010. See also *New Orleans Times-Picayune* 2010a; offshore-technology.com 2010.

3. See U.S. National Oceanic and Atmospheric Administration 2010; see also Freudenburg et al. 2009.

4. Breed and McGill 2010; see also offshore-technology.com 2010.

Chapter 1: A Question for Our Time

1. The *New York Times* article was by Banerjee 2002.

2. This compilation of quotations was first assembled as part of what Lubin 2010 calls "Mike Milken's Excellent Presentation on Our Pathetic History of Foreign Oil Dependence." See also Vega 2010.

Chapter 2: The Macondo Mess

1. *New Orleans Times-Picayune* 2010b; see also offshore-technology.com 2010.

2. Raines 2010a.

3. On the Census of Marine Life, see PLoS One 2010; on dangers to birds, see for example Minard 2010, Rioux 2010; on the effects of the Loop Current, see Biello 2010, Borenstein 2010.

4. Henry 2010; Harris 2010a; BBC News 2010; Polson 2010; Gramling and Freudenburg 1992b. The final estimates were reported by Deepwater Horizon Study Group 2010; see also Boxall 2010.

5. Beamish 2002.

6. The quotation is from Kindy 2010; see also Levin 2010.

7. Hamburger and Geiger 2010; the quotation is from p. A12.

8. Associated Press 2010; Boxall and Tankersley 2010.

9. For thoughtful analyses of the stresses created by the "corrosive community" that emerged in the wake of the Exxon Valdez, see Marshall et al. 2004; Picou et al. 2004. For a journalist's summary, see Mauer 2010. For figures on deaths and liens on the long-delayed checks, see Pitts 2009.

10. The comparisons are from U.S. Government Accountability Office 2007, and from Iledare and Olatubi 2006. Broader assessments are provided by Gramling and Freudenburg 2009; McGowan 2010; Geiger and Hamburger 2010.

11. Geiger and Hamburger 2010. See also Government Accountability Office 2007.

12. We are building here on earlier thoughts from Erikson 1976, and Freudenburg et al. 2009.

13. For the original statement, see Peter and Hull 1969. See also Erikson 1976; Freudenburg et al. 2009.

14. Mills 2010.

Chapter 3: Stored Sunlight and Its Risks

1. For a more detailed discussion, see, e.g., Hyne 1995.

2. The quotation is from Hyne 1995, p. 173.

3. Hyne 1995; Gramling 1996.

4. The quotation is from Kaldany 2006, p. 5; see also Quist-Arcton 2007.

5. The video of the interview is available at Wade 2010. The worker being quoted is identified in the video as Daniel Barron III, one of the survivors of the *Deepwater Horizon* explosion.

6. See especially Freudenburg 1992a; Busenberg 1999.

7. For statements of the original concept, and applications to the *Exxon Valdez*, see especially Freudenburg 1992a, 1992b. See also Clarke 1993, 1999. Official estimates put the size of the Exxon spill at 11 million gallons, the figure used here, but other calculations indicate that the total size of that spill might have been as much as 38 million gallons. Even the larger figure, however, would have been smaller than the quantity of oil lost from the BP blowout, and in all likelihood, the true figure will never be known.

8. The quotation is from Bartimus et al. 1989, pp. 1, 15; see also Church, 1989.

9. For a thoughtful analysis, see Pulver 2007; see also Schwartz 2006a. The quotation is from Lustgarten and Knutson 2010a.

10. See for example the assessment by Associated Press business writer, Wardell 2010.

11. For further details, see for example Schwartz 2006b; Lustgarten 2010; Hatcher 2010.

12. The quotation about the "draconian" nature of cost-cutting operations comes from ProPublica reporter Abrahm Lustgarten (2010). See also Lustgarten and Knutson 2010b.

13. The information and quotation on the OSHA fine are from Greenhouse 2010. The quotation and story on the "opportunities for more cost cutting," come from Clanton 2010.

14. See the analysis by Morris and Pell 2010.

15. See Lustgarten and Knutson 2010a, which is the source of the quotation, and Lustgarten and Knutson 2010b.

16. Bluestein and Baker 2010.

17. Griffin and Fitzpatrick 2010.

18. Cart et al. 2010; the quotation is from p. AA5.

19. Lin and Boxall 2010; the quotation is from p. AA1.

20. The article was by Urbina 2010.

21. The quotation comes from the *Washington Post* story (Hilzenrath 2010). On the downgrading by Moody's, see, e.g., Chang 2010.

22. The details are provided in Waxman and Stupak 2010.

23. The quotation is from Waxman and Stupak 2010, p. 2.

24. Waxman and Stupak 2010.

25. McClatchy 2010.

26. The article in CNN.Money.Com was by *Fortune* writer, Nelson Schwartz (2006b).

27. Papritz 1983.

28. The quotations come from Kravitz and Flaherty 2008 and Joffe-Walt and Kestenbaum 2010.

29. FoxNews.com 2010.

30. Clarke 1999.

31. BP 2009 (Regional Oil Spill Response Plan—Gulf of Mexico, Appendix H), p. BP-HZN-CEC 000533.

32. Raines 2010b.

33. See Raines 2010b; Mohr et al. 2010.

34. Associated Press 2010.

35. Boxall and Tankersley 2010. The quotations are from p. AA1.

36. The quotations come from the story—Kaufman et al. 2010.

37. Associated Press 2010.

38. See Joffe-Walt and Kestenbaum 2010.

39. Taylor 2010.

40. Kane and Yourish 2010.

41. For the apology to BP, and then the apology for the apology, see, e.g., Mason 2010. For information on campaign contributions, see Silver 2010.

42. Anderson and Kunzelman 2010.

43. Anderson 2010.

44. Eggen and Kindy 2010.

45. Shapiro 1987.

Chapter 4: Colonel of an Industry

1. U.S. National Park Service 2002. The quotation is from p. 1.

2. For more discussion, see, e.g., Solberg 1976.

3. Pees 2004.

4. Giddens 1975.

5. For the ASTM timeline, see Totten, n.d.

6. Pees 2004; see also Solberg 1976.

7. Martin and Gelber 1978.

8. Sampson 1975. The quotation, which is also a major theme of his book, is from page ix.

9. The information about oil prices comes from Sampson 1975; the quote is from p. 21. The quotation about the pension is from Ockershausen 1995.

10. Sampson 1975; the statement about the post office is on p. 19.

11. For a more detailed account, see Darrah 1972.

12. Gramling 1996. The quotation is from p. 15.

13. For more technical discussions of "overadaptation," see Gramling and Freudenburg 1990, 1992a; Freudenburg 1992c.

14. For a classic discussion of the issue of "efficiency," see Bateson 1972.

Chapter 5: Barons and Barrels

1. See, e.g., Cochran and Miller 1942.

2. For more discussion, see Cronon 1991.

3. For further details, see Clark 1987; Solberg 1976.

4. The best-known of the assessments of oil companies by "muckrakers" was the later, book-length treatment by Tarbell 1904. For information on McKinley's campaign finances, see Flynn 1932.

5. Solberg 1976; Clark 1987.

6. The anecdote is from Solberg 1976.

7. *Handbook of Texas Online*. Available at http://www.tsha.utexas.edu/handbook/online/articles/view/SS/dos3.htm (accessed 4 Sep. 2004).

8. For further discussion, see Nicholson 1941.

9. For further discussion, see Sampson 1975 Freudenburg and Gramling 1994a; Solberg 1976.

10. A "stripper" operation is one that seeks to "strip out" the last few barrels of oil remaining in a field. The quotation is available at http://www.spindletop.org/timeline/index.html (accessed 2 Sept. 2005).

11. The quotation is from Solberg 1976, p. 73.

12. For more details on energy use patterns of the time, see Johnson 1979. For the quotations, see Laumann and Knoke 1987; Engdahl 2004.

13. For further discussions, see Nash 1968, especially pp. 85–86; see also Feagin 1990; Laumann and Knoke 1987; Solberg 1976.

14. For further discussions of these earlier warnings, see Ridgeway 1982, p. 90; Sampson 1975; see also *Scientific American* 1919. For figures on rail and auto transportation, see Snell 1974; see also U.S. Department of Commerce, Bureau of the Census, n.d.

15. For further discussion, see for example McCartney 2008.

16. The quotation comes from Rosner and Markowitz 1985, p. 345.

17. Rosner and Markowitz 1985, see also Freudenburg et al. 2008.

18. Nash 1968.

19. For further details, see Solberg 1976; Engler 1961; Ghanem 1986; Gramling 1996. For the British government's involvement see Yergin 1991.

20. See U.S. Federal Trade Commission 1952; the report's title, *The International Petroleum Cartel*, is a reference to the cartel formed under the Pact of Achnacarry, not to the cartel that became better-known to later generations, namely the Organization of Petroleum-Exporting Countries. For a more detailed description of the process, see also Ghanem 1986.

Chapter 6: Off the Edge in All Directions

1. For more information, see Myrick 1988; Freudenburg and Gramling 1994a.

2. On the land withdrawal and the Caddo Lake development process more broadly see Lankford 1971; see also Forbes 1946. On the early development of blowout preventers, see Brantly 1971, particularly pp. 1290–1305.

3. For more details, see especially Lankford 1971.

4. For more discussion of these points, see Freudenburg et al. 2009.

5. The "organic ooze" reference is from Russell 1942, p. 78. The comparison to bearing grease was first provided by Gramling 1996.

6. For more extensive discussions, see Williams 1934; Gramling 1996; Freudenburg and Gramling 1994a.

7. For further details, see *Oil Weekly* Staff 1946a.

8. The quotation is from Freudenburg and Gramling 1994a, p. 45. For land-loss calculations, see Gagliano 1973; Barrett 1970. For a more extensive and up-to-date discussion of land loss issues in the region, see Freudenburg et al. 2009. For the count of canals, see Davis and Place 1983.

9. Steinhart and Steinhart 1972.

10. The quotation is from Barry 1993, p. 43.

11. *World Oil* Staff 1956.

12. For further discussion, see Cicin-Sain and Knecht 1987; see also Engler 1961, particularly pp. 86–95.

13. The quotation is from Solberg 1976, p. 164.

14. The quotation is from *World Oil* Staff 1951, p. 73.

15. *Oil Weekly* Staff 1946b.

16. For further discussion, see Kaufman 1978.

17. Solberg 1976.

18. The quotation is from Sutton 1976, p. 67.

19. The reference to the Reconstruction Finance Corporation is from Laumann and Knoke 1989, pp. 58–59. See also Engdahl 2004; Gramling 1996.

20. Jones 1981; Ferrier 1982.

21. The quotations are from AmericanHeritage.com, n.d.

22. U.S. Federal Housing Administration 1939, as quoted in Jackson 1987, pp. 202–209.

23. The St. Louis calculations and most quotations are from Jackson 1987, pp. 202–209. The quotation about "helping to denude St. Louis of its middle-class residents" is from Jackson 1980, p. 434. See also Kelly 1993.

24. The quotations are taken from a transcript of "Taken for a Ride," a film by Jim Klein and Martha Olson, available at http://www.culturechange.org/issue10/taken-for-a-ride.htm (accessed 22 June 2010). For further information on the antitrust prosecution and convictions, see also Motavalli 2001; Mankoff 1999; Snell 1974.

25. Mieszkowski 2007; Shoup 2005; Davidson and Dolnick 2002.

26. The quotations are from AmericanHeritage.com, n.d.

27. The quotation is from U.S. Federal Highway Administration, n.d.

Chapter 7: "Energy Independence"

1. The quotation is from Gibbons and Chandler 1981, p. 2.

2. On OPEC, see Ghanem 1986. On the 1959 embargo, see also Feagin 1985.

3. The quotation is from Feagin 1985, p. 451. On OPEC breaking the hold of the multinational oil companies, see for example Ghanem 1986, especially pp. 141–150. For original data on oil prices, see *Oil and Gas Journal* 1988; see also Darmstadter and Landsberg 1976; Freudenburg and Gramling 1994a.

4. Much of this discussion draws heavily from Gramling and Freudenburg 1992b; for more on this era, see also Cooper 1973; for the map see Thomas 1946.

5. For more on this process, see for example Berry 1975.

6. Again, for more details, see Berry 1975; Cooper 1973.

7. Berry 1975.

8. See Thomas 1946; the map is on p. 44.

9. Cooper 1973; Berry 1975; Roscoe 1977.

10. Berry 1975.

11. The language is from the National Environmental Policy Act of 1969; for further discussions, see Freudenburg 1986; Llewellyn and Freudenburg 1990.

12. As quoted in Coates 1991, p. 196; see also Berry 1975.

13. The newspaper article is quoted in Molotch et al. 2000, p. 804.

14. On the radicalizing nature of the spill, see Molotch 1970. For the story from the *Santa Barbara Morning Press*, see Molotch et al. 2000. For some of our own earlier discussion of the political potency of the spill, see Freudenburg and Gramling 1994a.

15. For a more extensive account of the Watergate scandal, see for example White 1975.

16. For more details, see Freudenburg and Gramling 1994a; Gould 1989; Wilson 1982.

17. The Governors' resolution is quoted in Wilson 1982, p. 77. On the Gulf of Alaska, see especially Wunnicke 1982.

18. For more detailed analyses of the amendments, see Krueger and Singer 1979; Jones et al. 1979; Freudenburg and Gramling 1994a, 1994b.

19. The quotation is from Little 2004. The other person to be so singled out was Anne Gorsuch—later, Anne Gorsuch Burford—whom the same president appointed at about the same time to head the U.S. Environmental Protection Agency.

Chapter 8: To Know Us Is to Love Us?

1. The crowd estimates are from Baron 1989. For further discussion, see Gould et al. 1991; Gramling and Freudenburg 1996.

2. Morgan City Historical Society 1960; Comeaux 1972; Davis 1990.

3. Stanley et al. 1994.

4. On sociodemographic correlates of environmental concern, see Van Liere and Dunlap 1980. For Louisiana statistics from 1940, see U.S. Department of Commerce 1940. For a broader discussion of the points being raised here, see Freudenburg and Gramling 1993; 1994a.

5. For further discussion, see Freudenburg 1992c.

6. The questioning of sanity was not an official policy of the Department of Interior, but expressions of the belief that opposition reflected ignorance, selfishness, or irrationality, were widespread in the agency at the time. For the official Louisiana position, see Groat 1981; for further discussion of the Spiral of Stereotypes, see especially Freudenburg and Gramling 1993.

7. The previous record had been set by sale 37, held in February 1975, when memories of the 1973–1974 oil embargo were still fresh and oil prices were both high and rising. For further details, see Freudenburg and Gramling 1994a.

8. Jones et al. 1979; Farrow 1990.

9. For further discussion, see Gramling 1996.

10. For the official report, see National Research Council 1989.

11. For the *National Enquirer* article, see Mullins 1981. Local reports were sometimes considerably more insightful, particularly those from James Edmunds, who edited a weekly newspaper, *The Times of Acadiana*.

12. For oil-price data, see *Oil and Gas Journal* 1988.

13. The quotation is from Finn 1992, p. 38.

14. The gaps in the figure reflect the fact that there were no sales in 1956, 1957, 1958, 1961, 1963, or 1965.

15. A lease is a three mile square block, which contains 5,760 acres. The data on lease sales is from U.S. Minerals Management Service 2009. For a more detailed assessment, see Gramling and Freudenburg 2009.

16. Specifically, the Relief act allowed lease holders to remove as much as 98.5 billion cubic feet of gas or 17.5 million barrels of oil on leases in water depths between 200 and 400 meters; 295.6 billion cubic feet of gas or 52.5 million barrels of oil on leases in water depths between 400 and 800 meters; and 492.6 billion cubic feet of gas or 87.5 million barrels of oil on leases in water depths between greater than 800 meters. The figures are from U.S. Energy Information Administration 2009 and U.S. Minerals Management Services 2009.

17. U.S. Government Accountability Office 2007:2.

18. As quoted in Kocieniewski 2010.

19. The treasury assessment is from Krueger 2009; he is drawing on a study by the U.S. Congressional Budget Office 2005.

20. The quotations come from Kocieniewski 2010.

Chapter 9: Cleaning Up

1. For the observations on the Santa Barbara oil spill, see Molotch 1970; Molotch and Lester 1975.

2. For the *Times-Picayune* study, see Rioux 2010. For independent scientists' estimates of the remaining contamination, see Borenstein 2010; see also Harris 2010b.

3. Clarke 1999.

4. Perrow 1999; see also Harvey 1970.

5. Davis 2010.

6. Kletz 2010.

7. Bea 1998.

8. Perrow 1999, p. 188.

9. See for example Hamburger 2010.

10. Bea 2006.

11. Deepwater Horizon Study Group 2010.

12. For more details, see Busenberg 1999, 2007.

13. For those who love such details, Section (4)(b)(1) of the Occupational Safety and Health Act (29 U.S.C. 653) preempts OSHA from enforcing its regulations if a working condition is regulated by another agency of the federal government: "Nothing in this Act shall apply to working conditions of employees with respect to which other Federal agencies, and State agencies acting under section 274 of the Atomic Energy Act of 1954, as amended (42 U.S.C. 2021), exercise statutory authority to prescribe or enforce standards or regulations affecting occupational safety or health."

14. Barab 2010.

Chapter 10: Today and Tomorrow

1. Groat 1981.
2. On OCS production, see especially Manuel 1984; Gramling 1996.
3. Bloomberg News 2010.

References

AmericanHeritage.com. n.d. The Greatest Public Works Program in the History of the World. Available at http://www.americanheritage.com/articles/web/20060629-dwight-eisenhower-interstates-federal-aid-highway-act-autobahn-superhighways-charles-keralt-lucius-clay-aaa.shtml (accessed 22 June 2010).

Anderson, Curt, and Michael Kunzelman. 2010. Judge Who Nixed Drilling Ban Has Oil Investments (23 June). Available at http://www.washingtontimes.com/news/2010/jun/23/judge-who-nixed-drilling-ban-has-oil-investments/ (accessed 27 June 2010).

Anderson, Curt. 2010. AP IMPACT: Many Gulf Federal Judges Have Oil Links, Complicating Assignments On Spill Lawsuits (6 June). Available at http://www.ap.org/oil_spill/Manygulf_61010.html (accessed 22 June 2010).

Associated Press. 2010. Inspections of Deepwater Horizon Fell Dramatically Before Fatal Explosion. AI.com (May 16). Available at http://blog.al.com/live/2010/05/inspections_of_deepwater_horiz.html (accessed 14 June 2010).

Banerjee, Neela. 2002. This Oil's Domestic, but It's Deep and It's Risky. *New York Times* (11 Aug.). Available at http://www.nytimes.com/2002/08/11/business/this-oil-s-domestic-but-it-s-deep-and-it-s-risky.html (accessed 3 Feb. 2010).

Barab, Jordan. 2010. Testimony of Jordan Barab, Deputy Assistant for the Occupational Safety and Health Administration U.S. Department of Labor, Before the Committee on Health Education, Labor, and Pensions, Subcommittee on Employment and Workplace Safety, United States Senate, June 10, 2010.

Baron, Elliot. 1989. Black Friday. *Solares Hill* (July):10–11.

Barrett, Barney. 1970. *Water Measurements of Coastal Louisiana*. New Orleans: Louisiana Wildlife and Fisheries Commission.

Barry, Dave. 1993. Surviving the Swamp of Doom. *Wisconsin State Journal* (November 28): 4G.

Barstow, David, Laura Dodd, James Glanz, Stephanie Saul, and Ian Urbina. 2010. Regulators Failed to Address Risks in Oil Rig Fail-Safe Device. *New York Times* (20 June). Available at http://www.nytimes.com/gwire/2010/06/16/16greenwire-chemical-security-advocates-see-new-opening-to-2345.html (accessed 26 June 2010).

Bartimus, Tad, Hal Spencer, David Foster and Scott McCartney. 1989. Greed, Neglect Primed Oil Spill. *St. Louis Post-Dispatch* (April 9): 1, 15.

Bateson, Gregory. 1972. *Steps to an Ecology of Mind*. New York: Ballantine.

BBC News. 2010. US military joins Gulf of Mexico oil spill effort. BBC News (29 April). Available at http://news.bbc.co.uk/2/hi/americas/8651624.stm (accessed 18 June 2010).

Bea, R. G. 1998. Real-Time Prevention of Drilling & Workover Blowouts: Managing Rapidly Developing Crises. *Journal of Drilling Engineering*, May:1–10.

Bea, Bob. 2006. Learning from Failures: Lessons from the Recent History of Failures of Engineered Systems. Unpublished paper. Center for Catastrophic Risk Management. University of California, Berkeley.

Beamish, Thomas D. 2002. *Silent Spill: the Organization of an Industrial Crisis*. Cambridge: MIT Press.

Berry, Mary Clay. 1975. *The Alaska Pipeline: The Politics of Oil and Native Land Claims*. Bloomington: Indiana University Press.

Biello, David. 2010. Where Will the Deepwater Horizon Oil End Up? The Short Answer Is Everywhere—the Sea Surface, Deep Waters, the Gulf Coast, in Deepwater Corals and Even as Far as the Arctic. *Scientific American* (19 May). Available at http://www.scientificamerican.com/article.cfm?id=where-will-the-deepwater-horizon-oil-end-up (accessed 23 June 2010).

Bloomberg News. 2010. BP Negotiating Sale of Alaska Oil Fields: The Company Is Reportedly Talking to a Houston Firm and May be Trying to Raise Cash to Pay for the Spill. *Los Angeles Times* (12 July): A8. Available at http://articles.latimes.com/2010/jul/12/nation/la-na-0712-oil-spill-bp-20100712 (accessed 12 July 2010).

Bluestein, Greg, and Mike Baker. 2010. AP Exclusive: Witness says BP took "shortcuts." NewsTimes.com (26 May). Available at http://www.newstimes.com/default/article/AP-Exclusive-Witness-says-BP-took-shortcuts-500813.php (accessed 21 June 2010).

Borenstein, Seth. 2010. Major Study Charts Long-lasting Oil Plume in Gulf. *Atlanta Journal-Constitution* (20 Aug.). Available at http://www.ajc.com/news/nation-world/major-study-charts-long-595511.html (accessed 21 Aug. 2010).

Boxall, Bettina. 2010. Oil Spill Size Near Upper Range of Earlier Estimates: As BP Prepares to Seal the Well for Good, Officials Say It Spewed More than 200 Million Gallons, by Far the Worst Such Disaster off a U.S. Coast. *Los Angeles Times* (3 Aug.): AA1-AA2. Available at http://www.latimes.com/news/nationworld/nation/la-na-oil-spill-20100803,0,1234917.story (accessed 28 Aug. 2010).

Boxall, Bettina, and Jim Tankersley. 2010. Oil Rig Missed Inspections, Records Show. *Los Angeles Times* (June 12): AA1, AA7. Available at http://www.latimes.com/news/nationworld/nation/la-na-oil-spill-20100612,0,907235.story (accessed 14 June 2010).

BP. 2000. Regional Oil Spill Response Plan—Gulf of Mexico, UPS-US-SW-GOM-HSE-DOC-00177-2. Houston, TX: BP. Available at http://www.propublica.org/documents/item/bp-regional-oil-spill-response-plan-for-gulf-of-mexico#document/p3 (accessed 4 Aug. 2010).

Brantly, J. E. 1971. *History of Oil Well Drilling*. Houston: Gulf Publishing Co.

Breed, Allen G., and Kevin McGill. 2010. Inferno on the Gulf: Witnesses recount Rig Blast. *Houma Today* (May 9). Available at http://www.houmatoday.com/article/20100509/ARTICLES/100509297 (accessed 28 May 2010).

Busenberg, George. 1999. The Evolution of Vigilance: Disasters, Sentinels and Policy Change. *Environmental Politics* 8 (4) (Winter): 90–109.

Busenberg, George. 2007. Citizen Participation and Collaborative Environmental Management in the Marine Oil Trade of Coastal Alaska. *Coastal Management* 35:239–253.

Cart, Julie, Rong-Gong Lin, II, and Bettina Boxall. 2010 BP well to stay sealed as storm moves in: As Tropical Storm Bonnie approaches, vessels are prepared for evacuation. Also, during an investigative hearing, BP is accused of cutting corners. *Los Angeles Times* (23 July): AA1, AA5. Available at http://www.latimes.com/news/nationworld/

nation/la-na-oil-spill-20100723,0,1300454.story (accessed 23 July 2010).

Chang, Sue. 2010. Moody's cuts Transocean to "Baa3." MarketWatch.com (18 Aug.). Available at http://www.marketwatch.com/story/moodys-cuts-transocean-to-baa3-2010-08-18-1312210 (accessed 20 Aug. 2010).

Church, George J. 1989. The Big Spill: Bred from Complacency, the *Valdez* Fiasco goes from Bad to Worse to Worst Possible. *Time* (April 10):38–41.

Cicin-Sain, Biliana, and Robert Knecht. 1987. Federalism Under Stress: The Case of Offshore Oil and California. In *Perspectives on Federalism,* ed. Harry Scheiber, 149–176. Berkeley: Institute of Governmental Studies, University of California.

Clanton, Brett. 2010. BP Sees Opportunities for more Cost Cutting. *Houston Chronicle* (2 Feb.). Available at http://www.mysanantonio.com/business/83393747.html (accessed 3 July 2010).

Clark, John G. 1987. *Energy and the Federal Government: Fossil Fuel Policies, 1900–1946*. Urbana: University of Illinois Press.

Clarke, Lee. 1993. The Disqualification Heuristic: When Do Organizations Misperceive Risk? *Research in Social Problems and Public Policy* 5:289–312.

Clarke, Lee. 1999. *Mission Improbable: Using Fantasy Documents to Tame Disaster*. Chicago: University of Chicago Press.

Coates, Peter A. 1991. *The Trans-Alaska Pipeline Controversy: Technology, Conservation, and the Frontier*. Bethlehem: Lehigh University Press.

Cochran, T.G., and William Miller. 1942. *The Age of Enterprise*. New York: Macmillan.

Comeaux, Malcolm L. 1972. *Atchafalaya Swamp Life: Settlement and Folk Occupations*. Baton Rouge: Louisiana State University Press.

Cooper, Bryan. 1973. *Alaska: The Last Frontier*. New York: William Morrow & Co.

Coy, Peter, and Stanley Reed. 2010. Oil-Rig Disaster Threatens Future of Offshore Drilling. Bloomberg.com (May 14). Available at http://www.bloomberg.com/news/2010-05-06/bp-s-deepwater-horizon-rig-disaster-threatens-future-of-offshore-drilling.html (accessed 6 June 2010).

Cronon, William. 1991. *Nature's Metropolis: Chicago and the Great West*. New York: W.W. Norton.

Darmstadter, J., and H. Landsberg. 1976. The Economic Background. In *The Oil Crisis,* ed. R. Vernon, 15–37. New York: Norton.

Darrah, William Culp. 1972. *Pithole, the Vanished City: A Story of the Early Days of the Petroleum Industry*. Gettysburg: Pennsylvania Historical Association.

Davidson, Michael, and Fay Dolnick. 2002. *Parking Standards*. Chicago: American Planning Association, Planning Advisory Service.

Davis, Donald W. 1990. Living on the Edge: Louisiana's marsh, Estuary and Barrier Island Population. *Louisiana Geological Survey* 40:147–160.

Davis, Donald W., and John H. Place. 1983. *The Oil and Gas Industry of Coastal Louisiana and Its Effect on Land Use and Socioeconomic Patterns*. Reston, Va.: U.S. Geological Survey, Open File Report 83-118.

Davis, J. Frank. 2010. Personal Communication, Email to Gramling. 7/8/2010.

Deepwater Horizon Study Group. 2010. *Progress Report 2. Berkeley, CA: Center for Catastrophic Risk Management*. Berkeley: University of California.

Edson, Gary M., Donald L. Olson, and Andrew J. Petty. 2000. *Georges Bank Petroleum Exploration. Minerals Management Service, 2000–2031*. New Orleans: OCS Report MMS.

Eggen, Dan, and Kimberly Kindy. 2010. Three of Every Four Oil and Gas Lobbyists Worked for Federal Government. *Washington Post* (22 July). Available at http://www.washingtonpost.com/wp-dyn/content/article/2010/07/21/AR2010072106468.html (accessed 24 July 2010).

Engdahl, F. William. 2004. *A Century of War: Anglo-American Oil Politics and the New World Order*. Ann Arbor, Mich.: Pluto Press.

Engler Robert. 1977. *The Brotherhood of Oil: Energy Policy and the Public Interest*. Chicago: University of Chicago Press.

Erikson, Kai T. 1976. *Everything in Its Path: The Destruction of Community in the Buffalo Creek Flood*. New York: Simon and Schuster.

Farrow, R. Scott. 1990. *Managing the Outer Continental Shelf Lands: Oceans of Controversy*. New York: Taylor and Francis.

Feagin, Joe R. 1985. The Global Context of Metropolitan Growth. *American Journal of Sociology* 90:1204–1230.

Feagin, Joe R. 1990. "Extractive Regions in Developed Countries: A Comparative Analysis of the Oil Capitals, Houston and Aberdeen." *Urban Affairs Quarterly* 25 (4) (June): 591–619.

Ferrier, R. W. 1982. *The History of the British Petroleum Company*. Cambridge: Cambridge University Press.

Finn, Kathy. 1992. Where Are They Now? New Orleans Area S&Ls. *New Orleans Magazine* 26 (4) (January): 38–39.

Flynn, John T. 1932. *God's Gold, the Story of Rockefeller and His Times*. New York: Harcourt Brace Jovanovich.

Forbes, Gerald. 1946. A History of the Caddo Oil and Gas Field. *Louisiana Historical Quarterly* 29:3–16.

FoxNews.com. 2010. Interior Department Postpones Safety Awards Luncheon That Included BP (3 May). Available at http://www.foxnews.com/politics/2010/05/03/interior-dept-postpones-luncheon-honoring-safety-measures-offshore-oil-gas/ (accessed 21 June 2010).

Freudenburg, William R. 1986. Social Impact Assessment. *Annual Review of Sociology* 12:451–478.

Freudenburg, William R. 1992a. Nothing Recedes Like Success? Risk Analysis and the Organizational Amplification of Risks. *Risk: Issues in Health and Safety* 3 (1) (Winter): 1–35.

Freudenburg, William R. 1992b. Heuristics, Biases, and the Not-So-General Publics: Expertise and Error in the Assessment of Risks. In *Theories of Risk*, ed. Dominic Golding and Sheldon Krimsky, 229–249. Westport, Conn.: Greenwood.

Freudenburg, William R. 1992c. Addictive Economies: Extractive Industries and Vulnerable Localities in a Changing World Economy. *Rural Sociology* 57 (3) (Fall): 305–332.

Freudenburg, William R. 2005. Privileged Access, Privileged Accounts: Toward a Socially Structured Theory of Resources and Discourses. *Social Forces* 94 (1):89–114.

Freudenburg, William R., and Margarita Alario. 2007. Weapons of Mass Distraction: Magicianship, Misdirection, and the Dark Side of Legitimation. *Sociological Forum* 22 (2) (June): 146–173.

Freudenburg, William R., and Robert Gramling. 1992. *The Perceived Risks of Offshore Oil and Gas Development: Results of a Pilot Study*. Herndon, Va.: U.S. Minerals Management Service.

Freudenburg, William R., and Robert Gramling. 1993. Socio-Environmental Factors in Resource Policy: Understanding Opposition and Support for Offshore Oil Development. *Sociological Forum* 8 (3) (September): 341–364.

Freudenburg, William R., and Robert Gramling. 1994a. *Oil in Troubled Waters: Perceptions, Politics, and the Battle over Offshore Oil.* Albany: State University of New York (SUNY) Press.

Freudenburg, William R., and Robert Gramling. 1994b. Bureaucratic Slippage and Failures of Agency Vigilance: The Case of the Environmental Studies Program. *Social Problems* 41 (2) (May): 214–239.

Freudenburg, William R, Robert Gramling, and Debra Davidson. 2008. Scientific Certainty Argumentation Methods (SCAMs): Science and the Politics of Doubt. *Sociological Inquiry* 78 (1):2–38.

Freudenburg, William R., Robert Gramling, Shirley Laska, and Kai T. Erikson. 2008. Organizing Hazards, Engineering Disasters? Improving the Recognition of Political-economic Factors in the Creation of Disasters. *Social Forces* 87:1015–1038.

Freudenburg, William R., Robert Gramling, Shirley Laska, and Kai Erikson. 2009. *Catastrophe in the Making: The Engineering of Katrina and the Disasters of Tomorrow*. Washington, D.C.: Island Press.

Gagliano, S. M. 1973. *Canals, Dredging, and Land Reclamation in the Louisiana Coastal Zone. Report 14*. Baton Rouge: Center for Wetland Resources, Louisiana State University.

Geiger, Kim, and Tom Hamburger. 2010. Oil Companies Have a Rich History of U.S. Subsidies. *Los Angeles Times* (25 May). Available at http://articles.latimes.com/2010/may/25/nation/la-na-oil-spill-subsidies-20100525 (accessed 20 June 2010).

Geller, Raymond. 2010. BP oil spill rate in Gulf may be 3 million gallons per day. *World News Examiner*, May 14.

Ghanem, Skukri. 1986. *OPEC: The Rise and Fall of an Exclusive Club*. London: Routledge and Kegan Paul.

Gibbons, John H., and William U. Chandler. 1981. *Energy: The Conservation Revolution*. New York: Plenum.

Giddens, Paul H. 1975. *Edwin L. Drake and the Birth of the Petroleum Industry*. Harrisburg: Pennsylvania Historic and Museum Commission, Historic Pennsylvania Leaflet No. 21.

Gould, Gregory J. 1989. *OCS National Compendium*. Washington, D.C.: U.S. Minerals Management Service.

Gould, Gregory J., Robert M. Karpas, and Douglas L. Slitor. 1991. *OCS National Compendium*. Herndon, Va.: Minerals Management Service.

Gramling, Robert. 1996. *Oil on the Edge: Offshore Development, Conflict, Gridlock*. New York: State University of New York Press.

Gramling, Robert, and William R. Freudenburg. 1990. A Closer Look at "Local Control": Communities, Commodities, and the Collapse of the Coast. *Rural Sociology* 55 (4):541–558.

Gramling, Robert, and William R. Freudenburg. 1992a. Opportunity-Threat, Development, and Adaptation: Toward a Comprehensive Framework for Social Impact Assessment. *Rural Sociology* 57 (2) (Summer): 216–234.

Gramling, Robert, and William R. Freudenburg. 1992b. The *Exxon Valdez* Oil Spill in the Context of U.S. Petroleum Politics. *Industrial Crisis Quarterly* 6 (3):175–196.

Gramling, Robert, and William R. Freudenburg. 1996. Crude, Coppertone®, and the Coast: Developmental Channelization and the Constraint of Alternative Development Opportunities. *Society & Natural Resources* 9:483–506.

Gramling, Robert, and William R. Freudenburg. 2006. Attitudes Toward Offshore Oil Development: A Summary of Current Evidence. *Ocean and Coastal Management* 49:442–461.

Gramling, Bob, and Bill Freudenburg. 2009. Pay, Baby, Pay: Before the U.S. Responds to 'Drill, Baby, Drill' Campaign Rhetoric with More Offshore Energy Exploration, it Should Revise Reagan-era Leasing and Royalty Rules That Cost the Treasury Billions. *Miller-McCune* 2 (4) (July/Aug.): 40–46.

Greenhouse, Steven. 2010. BP to Pay Record Fine for Refinery. *New York Times* (12 Aug). Available at http://www.nytimes.com/2010/08/13/business/13bp.html?scp=1&sq=BP%20to%20Pay%20Record%20Fine%20for%20Refinery&st=cse (accessed 3 September 2010)

Griffin, Drew, and David Fitzpatrick. 2010. BP Plans to Get Rid of Safety Watchdog, Sources Say. CNN.Com (1 July). Available at http://www.cnn.com/2010/US/06/30/gulf.bp.ombudsman/index.html?hpt=T2 (accessed 3 July 2010).

Groat, C. G. 1981. Letter from Charles G. Groat, director of the Louisiana Geological Survey, to John L. Rankin, U.S. Department of Interior, Bureau of Land Management, September 15, 1981.

Hamburger, Tom. 2010. Rig's Command Structure Scrutinized: The Coast Guard will Investigate Whether Having the Drilling Manager in Charge Caused Confusion on Deepwater Horizon. *Los Angeles Times*

(24 June). Available at http://articles.latimes.com/2010/jun/24/nation/la-na-oil-spill-hearing-20100624 (accessed 25 June 2010).

Hamburger, Tom, and Kim Geiger. 2010. Foreign Flagging of Offshore Rigs Skirts U.S. Safety Rules: The Marshall Islands, not the U.S., had the Main Responsibility for Safety Inspections on the Deepwater Horizon. *Los Angeles Times* (June 15): A12. Available at http://www.latimes.com/news/nationworld/nation/la-na-oil-inspection-20100615,0,3043517,full.story (accessed 16 June 2010).

Harris, Richard. 2010a. Gulf Spill May Far Exceed Official Estimates. National Public Radio (14 May). Available at. http://www.npr.org/templates/story/story.php?storyId=126809525 (accessed 21 June 2010).

Harris, Richard. 2010b. Scientists Find Thick Layer of Oil on Seafloor. Washington, D.C.: National Public Radio, All Things Considered (10 Sept.). Available at http://www.npr.org/templates/story/story.php?storyId=129782098 (accessed 11 Sept. 2010).

Harvey, Tad. 1970. The Bottomless Wonder: 20-Story Oil Tank Stores on Floor of Arabian Gulf. *Popular Science* (Jan.): 80-83. Available at http://www.popsci.com/archive-viewer?id=OwEAAAAAMBAJ&pg=80&query=dubai+oil (accessed 23 June 2010).

Hatcher, Monica. 2010. Texas City Residents Unaware of Release at BP Refinery. *Houston Chronicle* (18 Aug). Available at http://www.chron.com/disp/story.mpl/business/7159288.html (accessed 20 Aug. 2010).

Henry, Ray. 2010. Scientists up Estimate of Leaking Gulf Oil: Flow Could Be up to 2.5 Million Gallons a Day, According to Federal Panel. MSNBC.com (15 June). Available at http://www.msnbc.msn.com/id/37717335/ns/disaster_in_the_gulf/ (accessed 20 June 2010).

Hilzenrath, David S. 2010. Oil Rigs' Safety Net Questioned as Governments Rely on Private Inspections. *Washington Post* (15 Aug). Available at http://www.washingtonpost.com/wp-dyn/content/article/2010/08/13/AR2010081306641.html (accessed 16 Aug. 2010).

Hyne, Norman J. 1995. *Nontechnical Guide to Petroleum Geology, Exploration, Drilling and Production*. Tulsa, Okla.: Pennwell Publishing Company.

Iledare, Omowumi O., and Williams O. Olatubi. 2006. *Economic Effects of Petroleum Prices and Production in the Gulf of Mexico OCS on the U.S. Gulf Coast Economy*. New Orleans, La.: U.S. Minerals Management Service, OCS Study MMS 2006-063. Available at http://

www.gomr.mms.gov/PI/PDFImages/ESPIS/4/4195.pdf (accessed 22 June 2010).

Jakobsson, M., and T. Flodén, and the Expedition 302 Scientists. 2006. Expedition 302 Geophysics: Integrating past Data with New Results. In Backman, J., Moran, K., McInroy, D.B., Mayer, L.A., and the Expedition 302 Scientists, Proc. IODP, 302: Edinburgh (Integrated Ocean Drilling Program Management International, Inc.). doi:10.2204/iodp.proc.302.102.2006

Jackson, Kenneth T. 1987. *Crabgrass Frontier: The Suburbanization of the United States*. New York: Oxford University Press.

Jackson, Kenneth T. 1980. Federal Subsidy and the Suburban Dream: The First Quarter-Century of Government Intervention in the Housing Market. *Records of the Columbia Historical Society, Washington, D.C.* 50:421–451.

Joffe-Walt, Chana, and David Kestenbaum. 2010. Unhappy With A Government Agency? Change The Name! All Things Considered, National Public Radio (23 June). Available at http://www.npr.org/templates/transcript/transcript.php?storyId=128059441 (accessed 24 June 2010).

Johnson, James P. 1979. *The Politics of Soft Coal: The Bituminous Industry from World War I Through the New Deal*. Urbana: University of Illinois Press.

Jones, Geoffrey. 1981. *The State and the Emergence of the British Oil Industry*. London: Macmillan.

Jones, Russell O., Walter J. Mead, and Philip E. Sorenson. 1979. The Outer Continental Shelf Lands Act Amendments of 1978. *Natural Resources Journal* 19:885–908.

Kaldany, Rashad. 2006. Global Gas Flaring Reduction: A Time for Action! Keynote Speech, Global Forum on Flaring & Gas Utilization. Paris, France: December 13th, 2006. http://worldbank.org/html/fpd/ggfrforum06/kadany.pdf (accessed 11 July 2007).

Kane, Paul, and Karen Yourish. 2010. Congress Members Overseeing Firms Involved in Gulf Spill Held Oil, Gas Stock. *Washington Post* (17 June). Available at http://www.washingtonpost.com/wp-dyn/content/article/2010/06/16/AR2010061605369.html (accessed 22 June 2010).

Kaufman, Burton I. 1978. *The Oil Cartel Case: A Documentary Study of Antitrust Activity in the Cold War Era*. Westport, Conn.: Greenwood Press.

Kaufman, Marc, Carol D. Leonnig, and David Hilzenrath. 2010. MMS Investigations of Oil-rig Accidents Have History of Inconsistency. Washington Post (18 July). Available at http://www.washingtonpost.com/wp-dyn/content/article/2010/07/17/AR2010071702807.html (accessed 20 July 2010).

Kelly, Barbara M. 1993. *Expanding the American Dream: Building and Rebuilding Levittown*. Albany: State University of New York Press.

Kindy, Kimberly. 2010. Recovery Effort Falls Vastly Short of BP's Promises. *Washington Post* (6 July). Available at http://www.washingtonpost.com/wp-dyn/content/article/2010/07/05/AR2010070502937.html (accessed 8 July 2010).

Kletz, Trevor. 2010. Personal Communication with Gramling, email 7/8/10.

Kocieniewski, David. 2010. As Oil Industry Fights a Tax, It Reaps Subsidies. New York Times (3 July). Available at http://www.nytimes.com/2010/07/04/business/04bptax.html?_r=1&scp=1&sq=As%20Oil%20Industry%20Fights%20a%20Tax,%20It%20Reaps%20Subsidies&st=cse (accessed 3 Sept. 2010).

Kravitz, Derek, and Mary Pat Flaherty. 2008. Report Says Oil Agency Ran Amok Interior Dept. Inquiry Finds Sex, Corruption. *Washington Post* (11 Sept.). Available at http://www.washingtonpost.com/wp-dyn/content/article/2008/09/10/AR2008091001829.html (accessed 15 June 2010).

Krueger. Alan B. 2009. Statement of Alan B. Krueger (Assistant Secretary for Economic Policy and Chief Economist, US Department of Treasury) to the Senate Subcommittee on Energy, Natural Resources, and Infrastructure, TG-284 (9 Sept). Available at http://www.treas.gov/press/releases/tg284.htm (accessed 16 July 2010).

Kruger, Robert B., and Louis H. Singer. 1979. An Analysis of the Outer Continental Shelf Lands Act Amendments of 1978. *Natural Resources Journal* 19:909–927.

Lankford Raymond, L. 1971. Marine Drilling. In *History of Oil Well Drilling*, ed. J. E. Brantly, 1358–1444. Houston: Gulf Publishing Co.

Laumann, Edward O., and David Knoke. 1987. *The Organizational State: Social Choice in National Policy Domains*. Madison: University of Wisconsin Press.

Levin, Alan. 2010. BP Incident may be Top Peacetime Oil Spill: High Estimate has this Spill Surpassing '79 Gulf Disaster. *USA Today* (2 July): 8A.

Lin, Rong-Gong II, and Bettina Boxall. 2010. Oil Worker Describes Bypass of Key Alarms. *Los Angeles Times* (24 July): AA1, AA7. Available at http://latimesblogs.latimes.com/greenspace/2010/07/gulf-oil-spill-bypassing-safety-systems-was-pervasive-on-rig.html (accessed 3 Sept. 2010).

Little, Amanda. 2004. How Green Was the Gipper? A look back at Reagan's Environmental Record. *Grist* (10 June). Available at http://www.grist.org/article/griscom-reagan/ (accessed 28 June 2010).

Llewellyn, Lynn G., and William R. Freudenburg. 1990. Legal Requirements for Social Impact Assessments: Assessing the Social Science Fallout from Three Mile Island. *Society & Natural Resources* 2 (3):193–208.

Lubin, Gus. 2010. Mike Milken's Excellent Presentation On Our Pathetic History Of Foreign Oil Dependence. BusinessInsider.com. Available at http://www.businessinsider.com/look-who-failed-to-reduce-foreign-oil-dependence-2010-4 (accessed 18 June 2010).

Lustgarten, Abrahm. 2010. BP Had Other Problems in Years Leading to Gulf Spill. Propublica.org (30 April). Available at http://www.propublica.org/article/bp-had-other-problems-in-years-leading-to-gulf-spill (accessed 3 May 2010).

Lustgarten, Abrahm, and Ryan Knutson. 2010a. Reports at BP over Years Find History of Problems. *Washington Post* (8 June). Available at http://www.washingtonpost.com/wp-dyn/content/article/2010/06/07/AR2010060704826.html (accessed 23 June 2010).

Lustgarten, Abrahm, and Ryan Knutson. 2010b Years of Internal BP Probes Warned That Neglect Could Lead to Accidents. ProPublica.org (7 June) http://www.propublica.org/feature/years-of-internal-bp-probes-warned-that-neglect-could-lead-to-accidents http://www.propublica.org/article/years-of-internal-bp-probes-warned-that-neglect-could-lead-to-accidents (accessed 3 Sept. 2010).

Mankoff, Al. 1999. Revisiting the American Streetcar Scandal. *In Transition* 4 (Summer). Available at http://www.intransitionmag.org/archive_stories/streetcar_scandal.aspxl (accessed 3 Sept. 2010).

Manuel, David P. 1984. Trends in Louisiana OCS Activities. In *The Role of Outer Continental Shelf Activities in the Growth and Modification of Louisiana's Coastal Zone,* ed. R. Gramling and S. Brabant, 27–40. Lafayette: U.S. Department of Commerce/Louisiana Department of Natural Resources.

Marshall, Brent K., J. Steven Picou, and Jan Schlichtmann. 2004. Technological Disasters, Litigation Stress and the Use of Alternative Dispute Resolution Mechanisms. *Law & Policy* 26 (2):289–307.

Martin, Michael, and Leonard Gelber. 1978. *The Dictionary of American History*. New York: Dorset Press.

Mason, Melanie. 2010. Rep. Joe Barton Backs Down from BP Apology, "Shakedown" Remark. *Dallas Morning News* (18 June). Available at http://www.dallasnews.com/sharedcontent/dws/news/politics/topstories/stories/061810dnnatbartonbp.b0aaffc4.html (accessed 29 June 2010).

Mauer, Richard. 2010. Gulf Spill Victims Learn from Alaskans. *Anchorage Daily News* (12 June). Available at http://www.adn.com/2010/06/11/1319427/gulf-spill-victims-learn-from.html (accessed 21 June 2010).

McCartney, Laton. 2008. *The Teapot Dome Scandal: How Big Oil Bought the Harding White House and Tried to Steal the Country*. New York: Random.

McClatchy. 2010. Rig's Manager Says BP Tried to Skip Test, Changed Drilling Plan. Available at http://www.mcclatchydc.com/2010/05/27/94963/doomed-rigs-manager-says-bp-tried.html (accessed 4 June 2010).

McGowan, Elizabeth. 2010. Research Shows Federal Oil Leasing and Royalty Income a Raw Deal for Taxpayers: Oil Industry Controls Huge Swaths of Public Land at World's Cheapest Prices. SolveClimate.com (18 May). Available at http://solveclimate.com/blog/20100518/research-shows-federal-oil-leasing-and-royalty-income-raw-deal-taxpayers (accessed 28 May 2010).

Mead, W., A. Moseidjord, D. Mauraoka, and P. Sorensen. 1985. *Offshore Lands: Oil and Gas Leasing and Conservation on the Outer Continental Shelf*. San Francisco: Pacific Institute for Public Policy Research.

Mieszkowski, Katherine. 2007. We paved paradise: So Why Can't We Find Any Place to Park? Because Parking Is One of the Biggest Boondoggles—and Environmental Disasters—in Our Country. Salon.com (1 Oct.) Available at http://www.salon.com/news/feature/2007/10/01/parking/ (accessed 22 July 2010).

Milken Institute. 2010. Global Conference: Shaping the Future

Mills, Mark P. 2010. Notes from Underground: Oil: Money, Politics and Power in the 21st Century. *Wall Street Journal* (2 July): A13.

Minard, Anne. 2010. Oil in Gulf of Mexico Spells Disaster for Young Birds as Breeding Season Unfolds. *Scientific American* (18 May). Available at http://www.scientificamerican.com/article.cfm?id=oil-spill-impact-on-birds (accessed 20 July 2010).

Mohr, Holbrook, Justin Pritchard, and Tamara Lush. 2010. AP IMPACT: BP's Error-prone Spill Response Plans Overstate Preparedness, Understate Dangers. CanadianBusinessOnline.com Available at http://www.canadianbusiness.com/markets/headline_news/article.jsp?content=b3598172 (accessed 21 June 2010).

Molotch, Harvey. 1970. Oil in Santa Barbara and Power in America. *Sociological Inquiry* 40 (Winter):131–144.

Molotch, Harvey, William R. Freudenburg, and Krista Paulsen. 2000. History Repeats Itself, but How? City Character, Urban Tradition, and the Accomplishment of Place. *American Sociological Review* 65:791–823.

Molotch, Harvey, and Marilyn Lester. 1975. Accidental News: The Great Oil Spill as Local Occurrence and National Event. *American Journal of Sociology* 81 (2): 235–260.

Moran, K., J. Backman, and J. W. Farrell. 2006. Deepwater Drilling in the Arctic Ocean's Permanent Sea Ice. In *Proc. IODP,* ed. J. Backman, K. Moran, D. B. McInroy, L. A. Mayer, and the Expedition 302 Scientists, 302: Edinburgh (Integrated Ocean Drilling Program Management)

Morgan City Historical Society. 1960. *A History of Morgan City, Louisiana*. Morgan City: Morgan City Historical Society.

Morris, Jim, and M. B. Pell. 2010. Renegade Refiner: OSHA Says BP Has "Systemic Safety Problem": 97% of Worst Industry Violations Found at BP Refineries. Center for Public Integrity (18 May). Available at http://www.publicintegrity.org/articles/entry/2085/ (accessed 8 June 2010).

Motavalli, Jim. 2001. *Forward Drive: The Race to Build "Clean" Cars for the Future*. San Francisco: Sierra Club.

Mouawad, Jad. 2009. Obama Tries to Draw Up an Inclusive Energy Plan. *New York Times* (March 17): B1.

MSNBC. 2008. Exxon Mobil posts $40.6 billion annual profit: Oil giant breaks record for largest annual profit by a U.S. company. http://www.msnbc.msn.com/id/22949325/ (accessed 22 Dec. 2008).

Mullins, Joe. 1981. Sleepy Backwater Becomes . . . Boom Town, U.$.A. *National Enquirer* (October 20): 9.

Myrick, David F. 1988. Summerland: The First Decade. *Noticias* (Quarterly Magazine of the Santa Barbara Historical Society) 34 (4) (Winter): 65–111.

Nash, Gerald D. 1968. *United States Oil Policy 1890–1964: Business and Government in Twentieth Century America*. Pittsburgh: University of Pittsburgh Press.

National Research Council. 1989. *The Adequacy of Environmental Information for Outer Continental Shelf Oil and Gas Decisions: Florida and California*. Washington, D.C.: National Academy Press, National Academy of Sciences.

National Research Council. 2003. *Cumulative Environmental Effects of Oil and Gas Activities on Alaska's north Slope*. Washington, D.C.: National Academy Press, National Academy of Sciences.

New Orleans Times-Picayune. 2010a. Burning oil rig sinks into Gulf of Mexico, Coast Guard says. *New Orleans Times-Picayune* (April 22). Available at http://www.nola.com/news/index.ssf/2010/04/burning_oil_rig_sinks.html (accessed 26 April 2010).

New Orleans Times-Picayune. 2010b. Sunken Rig Not Leaking Crude Oil, Coast Guard Official Says. *New Orleans Times-Picayune* (April 23). Available at http://topics.nola.com/tag/oil%20rig%20explosion/index-oldest-2.html (accessed 26 April 2010).

Nicholson, Gordon B. 1941. Spindletop Discovery Marked Birth of Modern Oil Industry. *The Oil Weekly* (6 Oct.).

Ockershausen, Jane. 1995. The Valley that Changed the World: Visiting the Drake Well Museum. *Pennsylvania Heritage Magazine* 21 (#3, Summer). Available at http://www.portal.state.pa.us/portal/server.pt/community/trails_of_history/4287/drake_well_museum_%28ph%29/472390 (accessed 3 Sept. 2010).

Offshore-technology.com. 2010. Deepwater Horizon: A Timeline of Events. Available at http://www.offshore-technology.com/features/feature84446/ (accessed 18 June 2010).

Oil and Gas Journal. 1988. *Oil and Gas Journal Data Book*. Tulsa, Okla.: Pennwell Books.

Oil Weekly Staff. 1946a. U-Shaped Drilling Barge. *The Oil Weekly* (2 Sept): 40–41.

Oil Weekly Staff. 1946b. World Crude Oil Production, by Countries, by Year. *The Oil Weekly* (11 Feb.): 100–102.

Onion.com. 2010. Massive Flow Of Bullshit Continues To Gush From BP Headquarters (June 7). Available at http://www.theonion.com/

articles/massive-flow-of-bullshit-continues-to-gush-from-bp,17564/ (accessed 8 June 2010).

Papritz, Carew, ed. 1983. *100 Watts: the James Watt Memorial Cartoon Collection*. Auburn, Wash.: Khyber Press.

Pees, Samuel T. 2004. *Oil History*. Meadville, Penn.A: Petroleum History Institute.

Perrow, Charles. 1999. *Normal Accidents: Living with High-Risk Technologies*. Princeton: Princeton University Press.

Peter, Laurence J., and Raymond Hull. 1969. *The Peter Principle: Why Things Always Go Wrong*. New York: Morrow.

Picou, J. Steven, Brent K. Marshall, and Duane Gill. 2004. Disaster, Litigation and the Corrosive Community. *Social Forces* 82 (4):1493–1522.

Pitts, Byron. 2009. Exxon Valdez Oil Spill: 20 Years Later. CBS News (2 Feb.). Available at http://www.cbsnews.com/stories/2009/02/02/eveningnews/main4769329.shtml (accessed 12 May 2010).

PLoS (Public Library of Science) ONE. 2010. Marine Biodiversity and Biogeography: Regional Comparisons of Global Issues. PLoS Collections: http://dx.doi.org/10.1371/issue.pcol.v02.i09 (accessed 8 Aug. 2010).

Polson, Jim. 2010. BP Gulf Well Gushing as Much as 60,000 Barrels a Day. Bloomberg Businessweek (16 June). Available at. http://www.businessweek.com/news/2010-06-16/bp-gulf-well-gushing-as-much-as-60-000-barrels-a-day-update3-.html (accessed 21 June 2010).

Pulver, Simone. 2007. Making Sense of Corporate Environmentalism: An environmental Contestation Approach to Analyzing the Causes and Consequences of the Climate Change Policy Split in the Oil Industry. *Organization & Environment* 20 (1):44–83.

Quist-Arcton, Ofeibea. 2007. Gas Flaring Disrupts Life in Oil-Producing Niger Delta. Washington, D.C.: National Public Radio, Morning Edition (July 24). http://www.npr.org/templates/story/story.php?storyId=12175714 (accessed 25 July 2007).

Raines, Ben. 2010a. Leaked Report: Government fears Deepwater Horizon well could become Unchecked Gusher. *Mobile Press-Register* (30 April). Available at http://blog.al.com/live/2010/04/deepwater_horizon_secret_memo.html (accessed 16 June 2010).

Raines, Ben. 2010b. BP told feds it could handle oil spill 60 times larger than Deepwater Horizon. *Mobile Press-Register* (19 May).

Available at http://blog.al.com/live/2010/05/bp_told_feds_it_could_handle_o.html (accessed 22 May 2010).

Reuters. 2010. Document Shows BP Estimates Spill up to 100,000 BPD. (20 June). Available at http://abcnews.go.com/US/wireStory?id=10964694 (accessed 12 July 2010).

Ridgeway, James. 1992. *Powering Civilization*. New York: Pantheon.

Rioux, Paul. 2010. Oil Spill Plugged, but More Oiled Birds than Ever Are Being Found. New Orleans Times-Picayune (8 Aug). Available at http://www.nola.com/news/gulf-oil-spill/index.ssf/2010/08/oil_spill_plugged_but_more_oil.html (accessed 10 Aug. 2010).

Roscoe, James P. 1977. *800 Miles To Valdez: The Building of the Alaskan Pipeline*. Englewood Cliffs: Prentice-Hall.

Rosner, David and Gerald Markowitz. 1985. A "Gift of God"? The Public Health Controversy over Leaded Gasoline During the 1920s. *American Journal of Public Health* 75 (4) (April): 344–352.

Russell, Richard J. 1942. Flotant. *Geographical Review* 32 (1)(January): 74-98.

Sampson, Anthony. 1975. *The Seven Sisters: The Great Oil Companies and the World They Made*. New York: Viking.

Schwartz, Nelson. 2006a. Can BP Bounce Back? A Disastrous Leak. A Deadly Explosion. CEO John Browne Must Turn His Troubled Oil Giant Around, but Time Is Running Out. *Fortune* (16 Oct.) Available at http://money.cnn.com/magazines/fortune/fortune_archive/2006/10/16/8388595/index.htm?postversion=2006100210 (accessed 20 June 2010).

Schwartz, Nelson. 2006b. BP Was Warned: Interviews with Employees and a 2002 Letter Predicting "Catastrophe" Show That BP's Problems Should Have Come as No Surprise to Management. Money.CNN.com (2 Oct.). Available at http://money.cnn.com/2006/10/02/magazines/fortune/BP_leak_short.fortune/index.htm (accessed 20 June 2010).

Scientific American. 1919. Declining Supply of Motor Fuel. *Scientific American* (8) (March): 220.

Shapiro, Susan. 1987. The Social Control of Impersonal Trust. *American Journal of Sociology* 93 (3):623–658.

Sheppard, Kate. 2008. Requiem for a Moratorium: Big Oil and Enviros Agree: Surging Prices Are Nail in Coffin for Offshore-Drilling Ban. *Grist* http://www.grist.org/article/requiem-for-a-moratorium (accessed 3 Sept. 2010)

Shoup, Donald. 2005. *The High Cost of Free Parking*. Chicago: Planners Press.

Silmon, David R. 2006. *Elite Deviance*. Boston: Pearson.

Silver, Nate. 2010. A Top Corporate Donor to Barton Is Partner of BP on Deepwater Horizon. Fivethirtyeight.com (17 June). Available at http://www.fivethirtyeight.com/2010/06/top-donor-to-barton-is-partner-of-bp-on.html (accessed 21 June 2010).

Snell, Bradford C. 1974. *American Ground Transport: a Proposal for Restructuring the Automobile, Truck, Bus, and Rail Industries.* Presented to the Subcommittee on Antitrust and Monopoly of the Committee on the Judiciary, United States Senate. February 26, 1974. Washington, D.C.: U.S. Government Printing Office.

Solberg, Carl. 1976. *Oil Power*. New York: Mason.

Stanley, D. R., C. A. Wilson, and C. Cain. 1994. Hydro-Acoustic Assessment of Abundance and Behavior of Fish Associated with Oil and Gas Platforms of the Louisiana Coast. *Bulletin of Marine Science* 55:1353.

Steinhart, Carol E., and John S. Steinhart. 1972. *Blowout: A Case Study of the Santa Barbara Oil Spill.* North Scituate, Mass.: Duxbury Press.

Sutton, Antony C. 1976. *Wall Street and the Rise of Hitler*. Seal Beach, Calif.: '76 Press.

Tarbell, Ida M. 1904. *The History of the Standard Oil Company*. London: McClure, Phillips & CO.

Taylor, Marisa. 2010. Since Spill, Feds Have Given 27 Waivers to Oil Companies in Gulf. McClatchy Newspapers (7 May). Available at http://www.mcclatchydc.com/2010/05/07/93761/despite-spill-feds-still-giving.html (accessed 21 June 2010).

Thomas, Henry F. 1946. A Report on an Expedition to the Far North to Locate Rumored Seeps and Secure Samples. *The Oil Weekly* (4 Feb.): 39–48.

Totten, George E. n.d. InContext: A Timeline of Highlights from the Histories of ASTM Committee DO2 and the Petroleum Industry. Available at http://www.astm.org/COMMIT/D02/to1899_index.html (accessed 14 June 2010).

Urbina, Ian. 2010. Gulf Oil Rig's Owner Had Safety Issue at 3 Other Wells. *New York Times*, 8/5/2010. Available at http://www.msnbc.msn.com/id/38570559/from/RSS/ (Accessed 5 August 2010).

U.S. Congressional Budget Office. 2005. Taxing Capital Income: Effective Rates and Approaches to Reform. Congress of the United States. Congressional Budget Office.

U.S. Department of Commerce, Bureau of the Census. n.d. *Census of Population and Housing: 1920 Census*. Available at http://www.census.gov/prod/www/abs/decennial/1920.html (accessed 3 June 2010).

U.S. Department of Commerce, Bureau of the Census. 1940. *1940 Census of Population: Characteristics of the Population*. Washington, D.C.: U.S. Government Printing Office.

U.S. Energy Information Administration. 2008. World Proved Reserves of Oil and Natural Gas. http://www.eia.doe.gov/emeu/international/reserves.html (accessed 15 Dec. 2008).

U.S. Energy Information Administration. 2009. Outer Continental Shelf Deep Water Royalty Relief Act of 1995. http://www.eia.doe.gov/oil_gas/natural_gas/analysis_publications/ngmajorleg/continental.html (accessed 1 July 2009)

U.S. Federal Highway Administration. n.d. Dwight D. Eisenhower National System of Interstate and Defense Highways. Available at http://www.fhwa.dot.gov/programadmin/interstate.cfm (accessed 22 June 2010).

U.S. Federal Trade Commission. 1952. *The International Petroleum Cartel*. Washington, D.C.: U.S. Government Printing Office.

U.S. Government Accountability Office. 2007. *Oil and Gas Royalties: A Comparison of the Share of Revenue Received from Oil and Gas Production by the Federal Government and Other Resource Owners*. Washington, D.C.: U.S. Government Accountability Office (GAO-07-676R Oil and Gas Royalties).

U.S. Minerals Management Service. 2006. *Assessment of Undiscovered Technically Recoverable Oil and Gas Resources of the Nation's Outer Continental Shelf, 2006*., 2006-01b. Herndon, Va.: MMS Fact Sheet RED.

U.S. Minerals Management Service. 2007. *Pipeline Damage Assessment from Hurricanes Katrina and Rita in the Gulf of Mexico (Technical Report)*. Herndon, Va.: Minerals Management Service.

U.S. Minerals Management Service. 2008. Virginia Lease Sale 220. http://www.boemre.gov/offshore/220.htm (accessed 3 Sept. 2010)

U.S. Minerals Management Service. 2009. MMS detailed lease data. http://www.gomr.mms.gov/homepg/pubinfo/freeasci/leasing/freeleas.html (accessed 7 Jan. 2009)

U.S. National Oceanic and Atmospheric Administration. 2010. BP Oil Spill Incident Response (update of 21 June 2010). Available at http://response.restoration.noaa.gov/dwh.php?entry_id=809 (accessed 21 June 2010).

U.S. National Park Service. 2002. A New Bedford Whaling. Brochure GPO:2002 491-282/40364. Washington, D.C.: U.S. Government Printing Office. Available at http://www.nps.gov/nebe/brochure/park-brochure.pdf (accessed 9 July 2005).

U.S. Office of Technology Assessment. 1994. *Saving Energy in U.S. Transportation.* OTA-ETI-559. Washington, D.C.: U.S. Government Printing Office.

Van Liere, Kent D., and Riley E. Dunlap. 1980. The Social Bases of Environmental Concern: A Review of Hypotheses, Explanations and Empirical Evidence. *Public Opinion Quarterly* 44:181–197.

Vega, Rocky. 2010. A History of False Starts for U.S. Energy Independence. *Christian Science Monitor* (30 April). Available at http://www.csmonitor.com/Business/The-Daily-Reckoning/2010/0430/A-history-of-false-starts-for-US-energy-independence (accessed 4 Sept. 2010).

Wade, Jared. 2010. BP's "Pattern of Neglect and Corner-Cutting." RiskManagementMonitor.com (8 June). Available at http://www.riskmanagementmonitor.com/bps-pattern-of-neglect-and-corner-cutting/ (accessed 15 July 2010).

Wardell, Jane. 2010. Tony Hayward Boosted BP's Bottom Line, but Not Safety." *New Orleans Times-Picayune* (26 July). Available at http://www.nola.com/news/gulf-oil-spill/index.ssf/2010/07/tony_hayward_boosted_bps_botto.html (accessed 7 Aug. 2010).

Waxman, Henry, and the Committee Chair, and Bart Stupak, Chair of Subcommittee on Oversight and Investigation, U.S. House Committee on Energy and Commerce. 2010. Letter to Tony Hayward, Chief Executive Officer, BP PLC. Available at http://energycommerce.house.gov/documents/20100614/Hayward.BP.2010.6.14.pdf (accessed 14 June 2010).

White, Theodore H. 1975. *Breach of Faith: The Fall of Richard Nixon.* New York: Athenum.

Williams, Neil. 1934. Practicability of Drilling Unit on Barges Definitely Established in Lake Barre, Louisiana Tests. *Oil and Gas Journal,* May 31: 47.

Wilson, Edward. 1982. MAGCRC: A Classic Model for State/Federal Communication and Cooperation. In *The Politics of Offshore Oil,* ed. Joan Goldstein, 72–90. New York: Praeger.

World Oil Staff. 1951. U.S. Tidelands Grab Only the Beginning. *World Oil* (September): 73.

World Oil Staff. 1956. Why Louisiana's Offshore Prospects Look Better. *World Oil* (April): 163–66.

Wunnicke, Esther C. 1982. The Challenge of the Alaskan OCS. In *The Politics of Offshore Oil*, ed. Joan Goldstein, 91-102. New York: Praeger.

Yergin, Daniel. 1991. *The Prize: The Epic Quest for Oil, Money and Power*. New York: Free Press.

Index

Afghanistan, 1
Agnew, Spiro, 123
Airline industry, 162–163
Alabama, 59
Alaska, 138, 166–167, 184
  ARCO and, 118
  BP and, 118, 165
  Bush (George W.) and, 2
  coastal residence and, 133–134
  Deadhorse, 123
  energy independence and, 115–127
  Exxon Valdez and, 17 (*see also* Exxon Valdez oil spill)
  Gravel Amendment and, 123
  Hickel (Walter) and, 118–121
  highways and, 133–134
  Humble and, 118
  Klondike gold rush in, 116
  limited resources of, 175–176
  litigation by State of, 140
  logistics of drilling in, 116
  Native protests in, 117–119, 121
  North Slope, 33, 43, 115–117, 119, 174
  oil leases in, 116–119, 124–126
  oil seep discovery in, 116
  Outer Continental Shelf and, 141
  Point Barrow, 116–117
  Prince William Sound, 37, 154, 156
  Prudhoe Bay, 3, 43, 118–119, 174
  public domain and, 117
  U.S. Department of the Interior and, 117–118, 120–121, 123, 125–128
  Valdez, 36, 38–39, 122, 175
  Yukon River, 123
Alaska Marine Highway, 134
Alaska Native Claims Settlement Act (ANCSA), 121
Alaskan Federation of Natives, 117
Alaskan pipeline
  Prudhoe Bay and, 43
  rupture of, 41
  Trans-Alaskan Pipeline Authorization Act, 123
  Trans Alaska Pipeline System (TAPS), 3, 102, 118, 120, 122–123
  Valdez and, 36

Alyeska Pipeline Service Company, 39, 120, 123
Amerada-Hess, 118–119
"American oil," 64
American Planning Association, 109–110
Amoco Caldiz spill, 156
Anadarko Petroleum, 57–58, 165
Andrus, Cecil, 127
Andry, Albert III, x–xi
Anglo-American Petroleum Company, 78
Anglo-Dutch Shell Oil Company, 103
Anglo-Iranian Oil Company, 103, 106, 114
Anglo-Persian Oil Company, 90, 102–103
Angola, 18
Antitrust laws, 197n24
  Justice Department and, 109
  Sherman Antitrust Act and, 78–79, 84, 177
  Texas and, 80, 82
Arabian Light crude oil, 114
Arabian Peninsula, 103
Arab-Israeli war, 114
Arab Oil Embargo of 1973, 113–114, 143, 183
Archbold, John, 80
ARCO (Atlantic Richfield Company), 118
Arctic Circle, 115
Arctic Ocean, 2–3
Arctic Slope Native Association, 117
Area-wide leasing
  environmental issues and, 139–140, 144–149, 172–173, 186
  U.S. Department of the Interior and, 172–173
Aspheron Peninsula, 66
Associated Press, 43, 56, 59
ASTM (American Society for Testing and Materials) International, 66
Astor, John Jacob, 76
Atchafalaya River, 135
Atrophy of vigilance, 35–39, 41–42, 159
Automobiles, 20, 37, 143
  advantage of oil and, 72
  American "love affair" with, 107
  dependency on, 85–86
  Federal-Aid Highway Acts and, 110–111
  General Motors and, 88, 109–111
  Great Depression and, 106–107
  housing and, 108–110
  Model T Ford and, 79
  National Interstate and Defense Highways Act and, 111–112
  oil consumption and, 106–107
  required parking for, 109–110
  streetcar removal and, 109
  World War II era and, 107

Baku, 66
Bankruptcy, 17, 111
Barab, Jordan, 169–170
Barge-mounted draglines, 96
Barrel unit, 71
Barron, Daniel, III, 192n5
Barry, Dave, 97
Barton, Joe L., 19, 58, 60, 194n41

Bayous, 94, 97, 134–135
Bea, Bob, 162
Benzene, 41
"Beyond Petroleum" slogan, 40
Bird migration, 11–12
Bligh Reef, 36–37, 51
Blowout preventers, 15, 29, 45, 55, 93, 196n2
  clean-ups and, 159–161
  development of, 93
  explanation of, 29
  previous success of, 10–11
  Sepulvado (Ronald) and, 44
  shear ram and, 159–161, 164
Blowouts. *See also* Deepwater Horizon (drilling rig)
  backups and, 50
  cement bond log and, 49
  gushers, 11, 17, 31, 80–83, 116, 156
  Loop Current and, 11, 54
  policy for reducing, 158–170
  skimmers and, 14, 53–54, 154–157
  Spindletop and, 80–83
Bourg, Wes, xi
Boycotts, 106
BP, 2, 74. *See also* Deepwater Horizon (drilling rig)
  Alaskan oil and, 118, 165
  Anarko Petroleum and, 58
  as Anglo-Iranian Oil Company, 103, 106, 114
  atrophy of vigilance and, 35–39, 41–42, 159
  Barton's apology to, 19, 58, 60, 194n41
  Beyond Petroleum slogan and, 40
  as British Petroleum, 39–40
  Brown and, 40
  casing installation decisions of, 47–49
  casual approach to safety by, 34–61, 161–162
  cement bond log and, 49
  centralizers and, 48–49
  clean-ups and, 153, 155–166, 169
  drilling mud removal and, 50–51
  environmental issues and, xii (*see also* Environmental issues)
  exclusion (debarring) and, 43, 164–166
  fines of, 41–42
  Hayward (Tony) and, xii, 12, 40–42, 46, 58, 152
  House Committee on Energy and Commerce and, 46–47
  job cutting by, 40–41
  judicial branch and, 59
  Macondo project and, 9–20, 25, 29, 41–42, 49, 55, 57, 159, 164, 174, 188–189
  Marianas and, 33
  Oil Spill Response Plan of, 53, 157–158, 194n31
  public relations strategies of, 12–14, 39–40
  refocusing and, 164, 170
  reform commitments of, 40–41
  as Renegade Refiner, 42
  shear ram and, 159–160, 164
  Subcommittee on Oversight and Investigations and, 46–47
  Texas City refinery explosion and, 41, 44, 169
  toxic chemical release by, 41

Blowout preventers (cont.)
Transocean and, 33–34, 43, 45–46, 52–53, 58–59, 159, 165
Britain, 90, 102–103, 106, 196n19
Brower, Charles, 116
Browne, John, 40
Brunei, 18
Bullwinkle (oil platform code name), 9
Bureau of Land Management, 117–118, 138, 172
Bureau of Ocean Energy Management, Regulation and Enforcement (BOEMRE), 57, 167–168
Burning off natural gas, 14, 32–33
Bush, George H.W., 6, 141–142
Bush, George W., 2–3, 6, 53, 56, 142, 178

Caddo Lake field, 92, 92–94, 196n2
California
Fall, Albert, and, 116
Guadalupe, 13
Montecito, 124
Oakland, 110
Outer Continental Shelf and, 125–127
Santa Barbara channel and, 97, 124, 131
Santa Barbara oil spill and, 10, 91, 123–126, 153–154, 173, 198n14
Summerland, 91–92, 124
Tidelands Controversy and, 98–101
United States v. California and, 100
U.S. Minerals Management Service (MMS) and, 130–132
California State Mineral Leasing Act, 97
Campaign finances, 58–61, 78, 195n4
Canada, 40, 65, 122
Cancer, 41
Carbon dioxide, 32, 40
Cartels, 78, 90, 103, 196n20
Carter, Jimmy, 5, 125–127, 140, 143
Casing, 28–30, 45–51, 67
Catalytic converter, 89
Cement, 32, 59
bond logs and, 49
casing and, 28–29, 45, 48–50
centralizers and, 48–49
sealing with, 32, 43, 48
sediment and, 22
Census of Marine Life, 11
Center for American Progress, 150
Central Intelligence Agency (CIA), 106
Centralizers, 48–49
Chadman, John, 90
Chandeleur Islands, 94
Chandler, William U., 113
Cheney, Richard, 56
Chevron, 177
Chicago, 76
China, 66, 177–178
China National Offshore Oil Corporation (CNOOC), 177
Chumash people, 63
Civil War, 75–76
Clark, William (explorer), 76
Clark, William P., Jr., 139

Clarke, Lee, 53, 158, 193n7
Clay, Lucius, 110
Clean Development Mechanism, 32–33
Clean-up of oil spills, 13, 124
  blowout preventers and, 159–161
  BP and, 153, 155–166, 169
  Deepwater Horizon spill and, 155–156, 159–160, 163–167
  Exxon Valdez spill and, 154–157, 166
  fantasy documents and, 158
  kill operations and, 156–157, 161
  lack of technology for, 153–156
  management issues and, 158–164
  National Research Council and, 154
  policy recommendations for, 158–164
  recovery rates of, 156
  Santa Barbara spill and, 153–154
  shear ram and, 159–160, 164
  skimmers and, 14, 53, 154–157
  staging of, 154–155
  transparency and, 162–163
Cleveland, Grover, 78
Cleveland, Ohio, 75–77
CNN, 33, 44, 51
Coal, 71–72, 84, 86, 104, 181
Coalition forces, 2
Code names for oil platforms, 9
Cognac (oil platform code name), 9
Colorado, 18, 24, 116, 150
Competitive bidding, 87, 101, 146
Conflicts of interest. *See* Corruption and conflicts of interest
ConocoPhillips, 42
Consumption of oil
  American Way of Life and, 185–189
  automobiles and, 106–107
  depletion rates and, 85, 172–173
  during World War II, 105
  energy crisis and, 113, 122–123
  future and, 178–184
  Naval Petroleum Reserves and, 86–87
  oil shortages and, 85–86, 88, 115–116, 123, 143
  overadaptation and, 73
  remaining oil estimates and, 174–175
  stretching of resources and, 182–183
  United States' lead in, 84–85, 106, 179–180
  U.S. Fuel Administration and, 84–85
Containment efforts, 13–14, 39, 157
Cooper, Anderson, 33
Corruption and conflicts of interest
  bribed politicians and, 57–61 (*see also* Politics)
  depletion allowance and, 84
  judicial branch and, 59
  McKinley (William) and, 78–79
  National Recovery Administration and, 89–90

Corruption and conflicts of interest (cont.)
Pact of Achnacarry and, 90, 103, 196n20
reform and, 40, 77, 87, 164
regulators and, 59–60 (*see also* Regulation)
Standard Oil and, 78–79
Tidelands Controversy and, 98–101
U.S. Minerals Management Service (MMS) and, 51–57, 61
War Revenues Act and, 85
Cost-cutting, 15, 37, 42, 169, 193n12
Cox, Archibald, 125
Cracking (in oil refining), 71
Crawford, Truitt, 43
Cullinan, Joseph, 82
Curzon, Lord, 85
Cuttings, 30
Cyprus, 103

Dallas Morning News, 58
D'Arcy, William Knox, 102–103
Deepwater Horizon (drilling rig), ix–xiii, 6–7, 35, 80, 129, 192n5
casing installation decisions on, 47–49
casual approach to safety on, 34, 36, 41–60
cement bond log and, 49
clean-ups and, 155–156, 159–160, 163–167
containment efforts at spill of, 13–14
crew size of, x
deaths on, xi
disabled safety systems on, 45
drilling mud removal and, 50–51
drilling operations of, 33–36
estimating size of spill at, 12–14
explosion of, xi–xii, 7, 43, 47, 55
false alarms and, 45
gas pressure and, 33
Halliburton and, ix, 48, 59
location and, 33
Macondo project and, 9–20, 25, 29, 41–42, 49, 55, 57, 159, 164, 174, 188–189
as nightmare well, 47
physical description of, ix–x
policy recommendations from, 166–167, 174
public awareness and, 93
public relations strategies and, 12–14
record-setting status of, ix–x, 15, 33
recovery plans and, 13–14, 53
shear ram and, 159–160
sinking of, xii
surviving members of, 33–34
technology of, ix–x, 15, 19
thrusters of, x
Deepwater Horizon Study Group, 164
Deepwater Horizon United Command, 13
Deep Water Royalty Relief Act, 149, 200n16
Depletion allowance, 84
Derricks, 70, 80–81, 91, 95, 124, 132
Deterding, Henry, 90
Dinosaurs, 4, 7, 22, 73–74, 175, 180–182, 185

Draglines, 96
Drake, Edwin L., 25, 67–69, 71, 81–82, 102
Drake engine, 69
Drake's Folly, 67
"Drill, baby, drill!" slogan, 178
Drilling bits, 2, 27–29, 67–68
Drilling mud, 28–31, 50–51
Drilling rights. *See* Leases
Drill string, 28–30, 159–160
Drive pipe, 67
Drug use, 51–57
Dry holes, 31–32, 95, 98
Dubai, 160–161
DuPont, 88

Earth Day, xii, 10, 92, 120, 124, 131
Economic issues
  American Way of Life and, 185–189
  area-wide leasing and, 139–140, 144–149, 172–173, 186
  artificially cheap prices and, 183–184
  bankruptcy and, 17, 111
  boycotts and, 106
  budgets and, 33–34, 39, 150–151, 166, 168, 171
  competitive bidding and, 87, 101, 146
  conflicts of interest and, 51–61
  cost-cutting and, 15, 37, 42, 169, 193n12
  cutbacks and, 38
  delayed settlements and, 17
  depletion allowance and, 84
  depletion rates and, 172
  dry holes and, 31–32, 95, 98
  efficiency and, 5, 28, 73, 84, 88, 143, 151, 181–183
  Exclusive Economic Zones and, 2
  Executive Order 9633 and, 99
  extraction costs and, 24
  fines and, 41–42, 109, 162, 170, 193n13
  foreign oil and, 5–6, 90, 113–114, 121–122
  frontier regions and, 115, 123–129, 133, 144–145, 149, 173
  future and, 178–184
  Gilded Age and, 76
  golden parachutes and, 152
  Great Depression and, 89, 106–107
  Iraq and, 1–2, 4, 66, 71, 101–103, 114
  junk bond status and, 46
  Klondike gold rush and, 116
  lawsuits and, 17, 87, 120
  leases and, 18–19, 53, 55 (*see also* Leases)
  liens and, 17, 192n9
  loss of U.S. petroleum preeminence, 101–112
  monopolies and, 75, 77–78, 82, 109
  offshore wells and, 95, 98–99 (*see also* Offshore wells)
  oil barons and, 75–90
  oil consumption and, 2, 75, 106, 143, 174, 182–183, 186, 188
  oil production levels and, 67–68, 81, 83, 89–90, 102, 104, 107, 114–115, 145, 162, 173–174, 179
  oil prices and, 18–19, 39, 68, 89, 102, 143, 150–151, 179, 195n9, 198n3, 199n7

Economic issues (cont.)
OPEC and, 114–115, 123, 125
Outer Continental Shelf Deep Water Royalty Relief Act and, 19, 149
overadaptation and, 73
pipelines and, 71–72
profit concerns and, 18, 38, 40, 42, 73, 77, 81, 83, 144, 150–151, 169–170, 182
railroads and, 76
royalties and, 18–19, 52, 56, 59, 121, 148–150, 171, 186
sabotage and, 71
Sherman Antitrust Act and, 78–79, 84, 177
social multiplier effect and, 137, 166
speculation and, 66, 70, 81
Spindletop and, 80–81
subsidies and, 150–151, 182–184
sunk costs and, 95
taxes and, 18–19, 83–85, 111, 145–152
War Revenues Act and, 85
water depth and, x, 2, 6, 11, 19, 22, 25, 27, 30–33, 50, 57, 66, 68, 93, 116, 149
Ehrlich, Paul, 73
Einstein, Albert, 7
Eisenhower, Dwight D., 1, 100–101, 106, 110
Embargoes, 199n7
Arab Oil Embargo of 1973 and, 113–114, 143, 183
of 1959, 113, 198n2
production control and, 114
Project Independence and, 113
Emergency Response Team, 38–39
Endorfin (boat), ix–xiii
Energy crisis (of 1970s), 113, 122–123
Energy Efficiency, 151, 183
Carter policy and, 5
drill string and, 28
fuel, 143
organisms and, 73, 181–182
tetraethyl lead and, 88
U.S. Fuel Administration and, 84
Energy independence, 4, 20, 128, 180, 183
Alaska and, 115–127
American Way of Life and, 185–189
area-wide leasing and, 139–140, 144–149, 172, 186
changing administrations and, 5–6
energy crisis and, 113, 122–123
foreign oil and, 5–6, 90, 113–114, 121–122
loss of U.S. petroleum preeminence, 101–112
Nixon and, 5, 113, 115, 119, 122–123, 125, 172
oil shortages and, 85–86, 88, 115–116, 123, 143
Project Independence and, 113, 115, 123
solar energy and, 5
Environmental impact statements (EISs), 97, 120–122, 172
Environmental issues
Alaskan pipeline and, 120–122
area-wide leasing and, 139–140, 144–149, 172, 186

atrophy of vigilance and, 35–39, 41–42, 159
bayous and, 94, 97, 134–135
bird migration and, 11–12
Census of Marine Life and, 11
clean-ups and, 153–170
depletion rates and, 172
Earth Day and, xii, 10, 92, 120, 124, 131
Everglades and, 97
extraction and, 73, 82–83, 136, 153, 174, 177, 182, 185
flaring and, 32–33
frontier regions and, 115, 123–129, 133, 144–145, 149, 173
future and, 178–184
General Land Office and, 92–93
greenhouse gases and, 32–33, 40
Hawyard (Tony) on, xii
Loop Current and, 11–12, 54
Louisiana's awareness of, 136–137
marshes and, 94, 96–98, 133, 156
National Environmental Policy Act (NEPA) and, 120, 123
Obama on, xii
oil consumption and, 2–4, 75, 106, 143, 174, 182–183, 186, 188
Process Safety Management for Highly Hazardous Chemicals and, 168
recovery efforts and, 13–14, 53, 89
required parking and, 109–110
skimmers and, 14, 53–54, 154–157
Spindletop and, 81, 83
tidal exchange and, 97
timing of BP oil spill and, 11–12
toxic chemicals and, 41, 168
Vanity Fair and, 40
waste and, 3–4, 38, 83
Watt (James) and, 51, 54, 128–129, 137–138, 172–173
wetlands and, 12, 94, 96–97, 115
Environmental Protection Agency (EPA), U.S., 43
Euphrates River, 66
Everglades, 97
Exclusion, 164–166
Exclusive Economic Zones, 2
Executive Order 9633, 99
Exploratory wells, 32–33, 57, 95–96, 116
Explosions
Deepwater Horizon and, xi–xii, 7, 43, 47, 55
seismic technology and, 26
Texas City refinery, 41, 44, 169
Exxon, 79, 118
ExxonMobil, 79, 98
Exxon Valdez oil spill, 10, 13, 59, 122, 175, 192n9, 193n7
atrophy of vigilance and, 35–39
Bligh Reef and, 36–37, 51
casual approach to safety and, 60
cutbacks and, 38
delayed legal payments from, 17
environment of non-concern with, 36–39
policy recommendations from, 166
risk management and, 34–35

Exxon Valdez oil spill (cont.)
staged clean-up of, 154–157, 166
technology of, 36
volume of, 53

Fail-safe conditions, 15–16
Fall, Albert, 87–89, 116
False alarms, 45
Fantasy documents, 158
Federal-Aid Highway Acts, 110–111
Federal Housing Administration, 108
Federal Register, 138
Federal Trade Commission, 90, 196n20
Feldman, Martin, 59
Financial disclosure reports, 59
Fines, 41–42, 109, 162, 170, 193n13
Fish, ix–x, 132–133
Flaring, 14, 32–33
Floating hotels, 98
Florida, 11
energy independence and, 126
Everglades and, 97
judicial conflicts of interest and, 59
Loop Current and, 54
offshore experimentation and, 99, 130–133
Submerged Lands Act and, 101
U.S. Minerals Management Service (MMS) and, 130–131
Watt (James) and, 140–141
Flow Rate Technical Group, 12
Ford, Gerald, 5, 125–126
Foreign oil, 5–6, 90, 113–114, 121–122
Fossils, 4, 20–21, 72–74, 117, 149, 185
France, 103
Franklin, Benjamin, 183
Frontier regions for oil production
Alaska and, 115
energy independence and, 115, 123–129, 133, 144–145, 149
Nixon and, 122–123, 125
Fuel
coal, 71–72, 84, 86, 104, 181
energy independence and, 113–128
kerosene, 65, 79
military needs of, 84, 86–87, 116, 118, 122
oil's energy quality as, 71
tetraethyl lead and, 88–89
U.S. Fuel Administration and, 84–85

Gahdhi, Sima J., 150
Gasoline, 7, 18, 41, 104
Arab embargo and, 114–115
energy independence and, 114–115
oil barons and, 79, 82
tetraethyl lead and, 88–89
General Land Office, 92–93
General Motors, 88, 109–111
Geology
cost-cutting and, 50
Drake well and, 68
exploratory wells and, 32–33, 57, 95–96, 116
hard substrate and, 135
igneous rock and, 21
improved understanding of, 89
limited oil resource and, 144, 177, 180, 188

oil formation and, 21–25
rock oil and, 64–66, 176
sedimentary rock and, 21–26
Spindletop and, 79
Geophones, 26
Germany, 104–105
Gesner, Abraham, 65
Gibbons, John H., 113
Gilded Age, 76
Giliasso, Louis, 95
Giliasso (barge), 95–96
Gingrich revolution, 149
Global positioning technology, x
Golden parachutes, 152
Good Friday, 1989, 37
Gordon, Steve, 34
Gould, Jay, 76
"Government fears Deepwater Horizon well could become Unchecked Gusher" report, 11
Graham, Bob, 130
Gravel Amendment, 123
Great Depression, 89, 106–107
Great Plains, 24–25
Greenhouse gases, 32–33, 40
Groat, Chip, 172
Gulf of Mexico, 59, 115, 143, 158, 177
area-wide leasing and, 139–140, 144–149, 172, 186
bird migration and, 11–12
"Gulf war" in, 2
hurricanes and, 10–11
Loop Current and, 11–12, 54
Macondo project and, 9–20, 25, 29, 41–42, 49, 55, 57, 159, 164, 174, 188–189
moratoria in other offshore regions and, 142, 172–173, 178
moratorium on deepwater leasing in, 57, 59,
number of production facilities in, 171–172
Outer Continental Shelf (OCS) and, 101, 126–127, 131–134, 136, 172–173
pipelines in, 171
Gulf Oil, 82, 92
Gushers, 11, 17, 31, 80–83, 116, 156

Halliburton, ix, 48, 59
Handbook of Texas Online, The, 81, 83
Harding, Warren, 87, 89
Hard substrate, 135
Harrell, Jimmy, 50–51
Harriman, Edward Henry, 76
Harrison, Benjamin, 78
Hay, Mark, 45
Hayward, Tony, xii, 12, 40–42, 46–47, 58, 152
Heather (oil platform code name), 9
Hickel, Walter, 118–121
HMS *Bounty*, 37
Hogg, James Stephen, 80, 82
Holland, 103, 114
Holly (oil platform code name), 9
Home Oil, 118
Hoover, Herbert, 89
House Committee on Energy and Commerce, 46–47
Housing, 108–110, 187
Hughes, Howard, 82
Humble Oil, 118
Hunt, H.L., 82
Hurricane Andrew, 10–11
Hurricane Ferdinand, 10

Hurricane Ida, 33
Hurricane Ike, xi
Hurricane Ivan, 10
Hurricane Katrina, xi, 10, 184, 186
Hurricane Lily, 187
Hurricane Oil, 187–188
Hurricane Rita, xi, 10
Hussein, Saddam, 1, 13

Ickes, Harold, 99, 104–105
Igneous rock, 21
Illinois, 94
Interstate Commerce Law, 78
Inupiat, 3
Iran, 105
  Anglo-Iranian Oil Company, 103, 106, 114
  boycott of, 106
  nationalized holdings and, 114
  shah of, 106
Iraq
  coalition forces in, 2
  Euphrates River and, 66
  Hussein and, 1, 13
  invasion of, 1–2, 4
  Oassem and, 114
  oil and, 1–2, 4, 66, 71, 101–103, 114
  Powell (Colin) on, 2–3
Iraq Petroleum Company, 103
Ixtoc I oil spill, 13, 156

Japan, 104, 109, 177
Joe Griffin (ship), xii
Jordan, 103
Judicial branch, 59, 79
Junk bonds, 46

Kansas, 94
Kazakstan, 18
Kerosene, 65, 79
Kerr-McGee, 172
Kicks (from natural gas pressure), 30–31, 47–48, 55
Kier, Samuel, 64–66
"Kill" operations on oil blowout, 156, 161
Klondike gold rush, 116
Kuwait, 4, 13, 102–103, 114
Kyoto Protocol, 32–33

Lago Petroleum Company, 93
Lake Erie, 93
Lake Maracaibo oil field, 93–95
Lawsuits, 17, 87, 120
Leaded gasoline, 88–89
Leases, 55, 58, 70, 157, 200nn15,16
  Alaska and, 116–119, 124–126
  area-wide leasing policy and, 139–140, 144–149, 172, 186
  bribery and, 87–88, 116
  federal law and, 87
  first Federal sale, 139
  Louisiana and, x, 9–12, 25, 33, 53, 92, 98–101, 130, 132, 161, 165
  Magnolia Petroleum Company and, 98
  Outer Continental Shelf and, 19, 101, 149
  politics and, 138–149
  rates, by region, 18
  royalties and, 18–19, 52, 56, 59, 121, 148–150, 171, 186
  taxes and, 18–19
  Texas and, 98–99
  U.S. Congress and, 18–19
  U.S. GAO and, 18
Lebanon, 103
Legal issues, 1, 171

Alaskan Natives and, 117–119, 121
drilling rights and, 18 (*see also* Leases)
energy independence and, 123–124
Executive Order 9633 and, 99
federal ownership issues and, 99–101
international cooperation and, 103–104
lawsuits and, 17, 87, 120
marginal sea and, 99
offshore wells and, 96–100
oil barons and, 79–81, 84, 90
Outer Continental Shelf (OCS) and, 101, 125–127, 131–134, 136, 172–173
Submerged Lands Act and, 101
Summerland, California mob and, 124
United States v. California and, 100
Lewis, Meriwether, 76
Libya, 114
Liens, 17, 192n9
Lloyd's Register, 46, 52
Lobbyists, 60, 100, 110, 151, 168
Loop Current, 11–12, 54
Louisiana, 54, 150, 173, 199nn4,6
bayous and, 94, 97, 134–135
bust of, 174
Caddo Lake field and, 92–94, 196n2
coastal residence and, 133–134
Eugene Island, 98
exploratory/development wells in, 96–97
geographic differences from other regions, 133–135
Grand Isle, 155
Groat (Chip) and, 172
hard substrate and, 135
Hurricane Lily and, 187
judicial conflicts of interest and, 59
land prices and, 119
limited resources of, 175–176
marginal sea and, 99
marshes and, 94, 96–98, 133, 156
Morgan City, 98, 132
number of production facilities near, 171–172
oil leases in, x, 9–12, 25, 33, 53, 98–101, 130, 132, 161, 165
offshore wells and, 91–99, 171–172
Outer Continental Shelf and, 131–134, 136, 172
public education programs and, 137–138
shrimp industry and, 132
social characteristics of, historically, 135–138
tax revenue of, 18
Tidelands Controversy and, 98–101
U.S. Minerals Management Service (MMS) and, 130–137, 140, 142–143
Watt (James) and, 129, 172
Louisiana Geological Survey, 172
Louisiana Land Oil and Gas, 55–56
Lucas, Anthony, 79–80
Lutz, Peter, 54

Macondo project, 41–42, 49, 57, 164, 174, 188–189. *See also* Deepwater Horizon (drilling rig)
beginning of, 9–10
BP propaganda and, 12–14
clean-ups and, 159
drilling rights and, 18
lawsuits over, 17
One Hundred Years of Solitude and, 9–10, 188
origin of code name, 9–10
ownership of oil from, 18
riser length of, 29
safety and, 15–16
sediment and, 25
terrible timing of spill, 11–12
U.S. Minerals Management Service (MMS) and, 55–56
Magnolia Petroleum Company, 98
Marginal sea, 99
Marianas (drilling rig), 33
Marlin (oil platform code name), 9
Márquez, Gabriel García, 9, 188
Marshall Islands, 15
Marshes, 94, 96–98, 133, 156
Martinez, Bob, 130
Masters, Ian, 54–55
Maximum likelihood estimation, 35
McKinley, William, 78–79
McNutt, Marcia, 12
Mellon, Andrew, 82
Melville, Herman, 64
Mid-Atlantic Governors Coastal Resource Council, 126
Mineral Policy Act, 87
Mining industry, 67, 83–84, 87, 102
Mining Law, 87
Mississippi Canyon, x, 9–12, 25, 33, 53, 161, 165
Mississippi River, 25, 94, 135
Missouri River, 76
Mobil, 79, 98, 118
Mobile Press-Register, 11
Moby Dick (Melville), 64
Model T Ford, 79
Monopolies, 75, 77–78, 82, 109
Montana, 25
Moody's Investors Service, 46
Moratoria, 142, 172–173, 178
Alaska and, 117–119
Bush and, 142, 178
Gulf of Mexico and, 57, 59
Morton, Rogers, 120–122
Mossadeq, Mohammed, 106
Mozzafar al-Din Shah Qajar, Shah, 102
Muckrakers, 78, 195n4
Mud diving, 135

National Academy of Sciences, 129, 141, 142
National City Lines, 109
National Emphasis Program (NEP), 169
National Enquirer, 143
National Environmental Policy Act (NEPA), 120, 123
National Housing Act, 108
National Interstate and Defense Highways Act, 111–112
National Park Service, 64
National Public Radio, 12, 52
National Recovery Administration, 89–90
National Research Council, 129, 142, 154
Natural gas, 149, 151, 162, 166

advantages of, 71–72
energy independence and, 116–118, 126
extraction of, 24–27
flaring and, 14, 32–33
geology of, 21–25
Louisiana and, 130, 132
migration of, 25
offshore wells and, 91, 96, 101
pressure of, 30–31, 33, 55, 74, 80
politicians' investments in, 57–59
production wells and, 32
safety and, 43, 45, 47–50
Naval Petroleum Reserves, 86–87, 116, 118
Netherlands, 103, 114
New Orleans Magazine, 144
New York City, 76
New York Times, 1–2, 46, 151
Nixon, Richard M.
Alaskan oil and, 115, 119, 122–123, 125
energy independence and, 5, 113, 115, 119, 122–123, 125, 172
Hickel (Walter) and, 119
offshore wells and, 115, 172
Project Independence and, 113, 115, 123
Watergate and, 125, 198n15
Nobel family, 77–78
North Atlantic, 93
North Dakota, 25
North Sea, 93
Norway, 18

Oassem, General, 114
Obama, Barack, xii, 6, 19, 53, 56–59, 147, 150, 178
Occupational Safety and Health Administration (OSHA), 41–42, 167–170, 193n13, 201n13
Odyssey spill, 156
Offshore wells
barges and, 93–96
bayous and, 94, 97, 134–135
blow-out preventers and, 10–11, 15, 29, 44–45, 55, 93, 159–161, 164
Caddo Lake field and, 92–94, 196n2
decayed vegetation and, 95
draglines and, 96
Eisenhower (Dwight) and, 1
Executive Order 9633 and, 99
federal ownership issues and, 99–101
floating hotels and, 98
Ickes (Harold) and, 99, 104–105
Lake Maracaibo field and, 93–95
legal issues and, 96–100
Louisiana and, 91–99, 171–172
marginal sea and, 99
Mississippi Canyon block 252 and, 9, 25, 33, 53, 161, 165
Nixon (Richard) and, 115, 123, 125, 172
pilings and, 92–95, 98
power supply for, 93–94
pressure issues and, 92–93
salt domes and, 92, 94
shear ram and, 159–160, 164
shipworm (*terredo*) and, 93
steel decks and, 93
Stevenson (Adlai) and, 100
Submerged Lands Act and, 101

Offshore wells (cont.)
sunk costs and, 95
technology for, 91–96
tidal exchange and, 97
Tidelands Controversy and, 98–101
United States v. California and, 100
Venezuela and, 91–95
Oil
clean-up efforts and, 13–14, 124, 153–154, 157
consumption of, 2–4, 75, 106, 143, 174, 182–183, 186, 188
containment efforts and, 13–14, 39, 157
crude, 11, 13, 37, 41, 53, 66, 68, 70–71, 89, 102–103, 113–114, 173
current U.S. reserves of, 180
depletion rates and, 85, 172–173
"destruction" of, 4
dinosaur era and, 4, 7, 22, 73–74, 175, 180–182, 185
disadvantages of, 72–73
discovering, 25–27
drilling techniques and, 27–33
early techniques for obtaining, 66–74
early U.S. and, 63–66
energy quality of, 71
estimate of remaining amounts, 174–175
Exclusive Economic Zones and, 2
foreign, 5–6, 90, 113–114, 121–122
as fossil fuel, 4, 20–21, 72–74, 117, 149, 185
future and, 178–184
geological perspective on, 21–25
Great Plains and, 24–25
igneous rock and, 21
Iraq and, 1–2, 4, 66, 71, 101–103, 114
kerosene and, 65, 79
as limited resource, 175–178
as medicine, 64–65
military needs of, 84, 86–87, 116, 118, 122
new sources of, 4–5
organic matter and, 22–23
origins of, 21–25
overadaptation to, 73
photosynthesis and, 21
as power (geopolitically), 105–106
production levels of, 67–68, 81, 83, 89–90, 102, 104, 107, 114–115, 145, 162, 173–174, 179
Project Independence and, 113, 115, 123
recovery of, 13–14, 53, 89
reservoirs and, 23–25, 30–33, 53, 80, 92, 175
as rock oil, 64–66, 176
sandstone and, 22
sedimentary rock and, 21–26
seeps and, 25, 63–67, 116
shales and, 22, 24
shortages of, 85–86, 88, 115–116, 123, 143
source rocks and, 23
as stored sunlight, 21–22
submersible rigs and, x, 95–96
technology and, ix–x (*see also* Technology)
unexplored territories for, 176–177

variations in, 23
viscosity and, 23, 31
voids and, 22
wasting of, 3–4
water depth and, x, 2, 6, 11, 19, 22, 25, 27, 30–33, 50, 57, 66, 68, 93, 116, 149
as World War II factor, 104–105
Oil City, 70
Oil Creek, 25, 77
Oil Division, U.S. Fuel Administration, 84–85
Oil drilling
blow-out preventers and, 10–11, 15, 29, 44–45, 55, 93, 159–161, 164
casing and, 28–30, 45–51, 67
cement use and, 22, 28–29, 32, 42–45, 48–50, 59
centralizers and, 48–49
clean-up and, 153–170
code names for, 9
cost-cutting and, 15, 37, 42, 169, 193n12
cuttings and, 30
drilling bits and, 2, 27–29, 67–68
drilling mud and, 28–31, 50–51
drill string and, 28–30, 159–160
dry holes and, 31–32, 95, 98
exploratory wells and, 32–33, 57, 95–96, 116
kicks and, 30–31, 47–48, 55
Macondo project and, 9–20, 25, 29, 41–42, 49, 55, 57, 159, 164, 174, 188–189
moratoria and, 19, 57, 117–119, 142, 178
policy recommendations for, 158–170
production wells and, 32, 59
risers and, 29
rocking horse pumps and, 31
sealing and, 29, 32, 43, 48, 54, 63, 159
shear ram and, 159–160, 164
stripper operations and, 83, 195n10
U.S. Minerals Management Service and, 16, 48, 51–52, 55, 61, 129–130, 139, 141, 167, 200nn15,16
Oil industry
Alaskan oil and, 115–127
anticompetitive activities and, 78–90, 103, 196n20
area-wide leasing and, 139–140, 144–149, 172–173, 186
barons of, 75–90
beginnings of, 66–74
cartels and, 78, 90, 196n20
China and, 177–178
clean-up and, 153–170
damage to U.S. economy from, 171
depletion allowance and, 84
Drake and, 25, 67–69, 71, 81–82, 102
frontier regions and, 115, 123–129, 133, 144–145, 149, 173
golden parachutes and, 152
international competition and, 102–112
Interstate Commerce Law and, 78
leases and, 18 (*see also* Leases)
lobbyists and, 60, 100, 110, 151, 168

Oil industry (cont.)
  Meadville, Pennsylvania and, 67–68
  monopolies and, 75, 77–78, 82, 109
  muckrakers and, 78, 195n4
  OPEC and, 64, 105, 114–115, 123, 125, 183, 198n2
  Pact of Achnacarry and, 90, 103, 196n20
  patriotism and, 104–105
  Peter Principle and, 19–20, 164
  Pithole, Pennsylvania and, 69–74, 91, 172–175
  policy recommendations for, 158–170
  Rockefeller (John) and, 75–84, 102
  royalties and, 18–19, 52, 56, 59, 121, 148–150, 171, 186
  Sherman Antitrust Act and, 78–79, 84, 177
  social multiplier effect and, 137, 166
  Spindletop and, 79–83, 91–92, 95, 102, 116, 172–173
  subsidies and, 150–151, 182–184
  Tidelands Controversy and, 98–101
  Titusville, Pennsylvania and, 66–67
  vertically integrated entities and, 82
  War Revenues Act and, 85
  wildcatters and, 4, 82–83, 152
Oil leases. *See* Leases
Oil prices
  artificially low, 183–184
  busts in, 68, 69–74, 81, 89–90, 143, 173–74
  economic issues and, 18–19, 39, 68, 89, 102, 143, 150–151, 178–179, 195n9, 198n3, 199n7
  embargoes and, 114 (*see also* Embargoes)
  engineered cuts in, 113–114
  National Recovery Administration and, 89–90
  new oil field discoveries and, 68, 80–83, 89
  Project Independence and, 113, 115, 123
  quadrupling of, 114–115
  Spindletop and, 81
Oil-rush cities
  Oil City, 70
  Oleopolis, 70
  Pithole, 69–74
  Spindletop, 81, 83
  Titusville, 66–69
Oil shale, 24
Oil Spill Response Plan, 53, 157, 194n31
Oil Weekly journal, 116, 118, 122
Oklahoma, 25, 89, 150
Oleopolis, 70
One Hundred Years of Solitude (García Márquez), 9, 188
Operation Ajax, 106
Oregon, 140
Organic matter, 22–23
Organization of Petroleum-Exporting Countries (OPEC), 64, 105, 114–115, 123, 125, 183, 198n2
Outer Continental Shelf (OCS), 101, 125–127, 173
  area-wide leasing and, 139–140, 144–149, 172, 186

Louisiana and, 131–134, 136, 172
Outer Continental Shelf Deep Water Royalty Relief Act, 19, 149
Outer Continental Shelf Lands Act (OCSLA), 101
Outer Continental Shelf Lands Act Amendments (OCSLAA), 127
Overadaptation, 73, 180–183
Overdesign (as safety measure), 73

Pact of Achnacarry, 90, 103, 196n20
Panama, 95
Panama Canal, 86
Parking Standards, 109–110
Patents, 65, 87, 95
Pearl Harbor, 104
Pennsylvania, 25, 27, 65
Meadville, 67–68
Pithole, 69–74, 76, 81, 83, 91, 172–173, 175
Pittsburgh, 76
Titusville, 66–69
Pennsylvania Rock Oil Co., 66–67
Perrow, Charles, 162
Persia, 102
Peter Principle, 19–20, 164
Petroleum Coordinator for National Defense, 105
Phillips (oil company), 118
Pilings, 92–95, 98
Pipelines
Alaskan, 3, 36, 39, 41, 43, 102, 118, 118–125
corridors for, 96
economic advantages of, 71–72
gas, 32
Gravel Amendment and, 123
miles of in Gulf of Mexico, 171–172
at Pithole, Pennsylvania 71
Reconstruction Finance Corporation and, 105, 197n19
sabotage of, 71
Spindletop and, 81
Thomas map and, 116, 122, 198n4
underwater, 92
world's first oil, 70–72
Policy
artificially cheap prices and, 183–184
damage to U.S. economy from, 171, 175
energy efficiency and, 5, 28, 73, 84, 88, 143, 151, 181–183
exclusion and, 164–166
extraction and, 73, 82–83, 136, 153, 174, 177, 182, 185
future and, 178–184
need for new technologies and, 183
refocusing and, 164, 170
reform of, 158–160
Regional Citizens' Advisory Councils and, 166
stretching of resources and, 182–183
transparency and, 162–163
Politics
area-wide leasing and, 172–173
Barton (Joe) apology and, 19, 58, 60, 194n41
Bush (George H.W.) and, 6, 141–142

Politics (cont.)
Bush (George W.) and, 2–3, 6, 53, 56, 142, 178
Carter and, 5, 125–127, 140, 143
clean-up and, 158–164
Cleveland (Grover) and, 78
conflicts of interest and, 51–61
drilling rights and, 18
Eisenhower (Dwight) and, 1, 100–101, 106, 110
energy independence and, 4–6, 20, 113–128, 144, 180, 183
Energy Plan and, 56
Exclusive Economic Zones and, 2
Executive Order 9633 and, 99
Federal-Aid Highway Acts and, 110–111
Ford (Gerald) and, 5, 125–126
General Motors and, 88, 109–111
Gingrich (Newt) revolution and, 149
Harding (Warren) and, 87, 89
Harrison (Benjamin) and, 78
Hickel (Walter) and, 118–121
Hogg (James) and, 80, 82
Hoover (Herbert) and, 89
Ickes (Harold) and, 99, 104–105
judicial branch and, 59
leases and, 138–149
lobbyists and, 60, 100, 110, 151, 168
McKinley (William) and, 78
muckrakers and, 78, 195n4
National Interstate and Defense Highways Act and, 111–112
Naval Petroleum Reserves and, 86–87
Nixon (Richard) and, 5, 113, 115, 119, 122–123, 125, 172
Obama (Barack) and, xii, 6, 19, 53, 56–59, 147, 150, 178
oil embargoes and, 90, 113–115, 123, 125, 143, 183, 198n2, 199n7
OPEC and, 64, 105, 114–115, 123, 125, 183, 198n2
Outer Continental Shelf Deep Water Royalty Relief Act and, 19, 149
Pact of Achnacarry and, 90, 103, 196n20
Peter Principle and, 19–20
Progressive Era and, 87–88
Qaddafi (Muammar) and, 114
Reagan (Ronald) and, 5–6, 51, 127–128, 136, 138–139, 143–144
Roosevelt (Franklin) and, 99
Roosevelt (Theodore) and, 79, 86–87
reform and, 40, 77, 87, 164
Stevenson (Adlai) and, 100
Taft (William) and, 79, 86–87
taxes and, 18–19, 83–85, 111, 145–152
Truman (Harry) and, 99
U.S. Congress and, 18–19 (*see also* U.S. Congress)
War Revenues Act and, 85
Watergate and, 125, 198n15
Watt (James) and, 51, 54, 128–129, 137–138, 172–173
Wilson (Woodrow) and, 84
Powell, Colin, 2–3
Process Safety Management for Highly Hazardous Chemicals, 168
Produced waters, 32

Production levels, 5, 12–14, 145, 157–158, 162
current U.S., 179
development of offshore methods and, 102, 104, 107
energy independence and, 114–115
extraction and, 73, 82–83, 136, 153, 174, 177, 182, 185
Lake Maracaibo field and, 93
oil barons and, 77, 81
policy issues and, 173–174, 179
Project Independence and, 115
Prudhoe Bay and, 174
quotas for, 89–90
Spindletop and, 83
U.S. Fuel Administration and, 84–85
Production wells, 32, 59
Profits, 18, 81, 83, 182
area-wide leasing and, 139–140, 144–149, 172, 186
BP and, 40, 42, 73
fines and, 170
frontier regions and, 115, 123–129, 133, 144–145, 149, 173
future and, 178–184
maximization of, 169
risk management and, 38
royalties and, 18–19, 52, 56, 59, 121, 148–150, 171, 186
Spindletop and, 81
Standard Oil and, 77
subsidies and, 150–151, 182–184
tax breaks and, 150
Progressive Era, 87–88
Project Independence, 113, 115, 123
ProPublica, 42

Qaddafi, Colonel Muammar, 114
Quotas, 89–90
Qutar, 114

Railroads, 71, 95, 109
Chicago and, 76
dependency on, 86
Interstate Commerce Law and, 78
modernizing of, 109
Rockefeller (John) and the outsmarting of, 76–80, 83
Standard Oil and, 79–80
Texas Railroad Commission and, 83
Rankin, John L., 172
Razmara, Ali, 106
Reagan, Ronald, 5–6, 51, 127–128, 136, 138–139, 143–144
Reconstruction Finance Corporation, 105, 197n19
Records of the Columbia Historical Society, 108
Refineries, 72
cracking and, 71
efficiency and, 71
first commercial, 65
Rockefeller (John) and, 75–77, 82
safety and, 41–45, 168–169
Texas City explosion and, 41, 44, 45, 169
Refocusing, 164, 170
Reform
Hayward (Tony) and, 40
policy recommendations for, 158–170
Progressive Era and, 87
Standard Oil Trust and, 77

Regional Citizens' Advisory Councils, 166
Regulation. *See also* Safety
- Bureau of Ocean Energy Management, Regulation and Enforcement and, 57, 167–168
- California State Mineral Leasing Act and, 97
- casual approach to safety and, 41–51
- corruption and conflicts of interest in, 51–61, 78–79
- depletion allowance and, 84
- Emergency Response Team and, 38–39
- Federal-Aid Highway Acts and, 110–111
- Federal Trade Commission and, 90, 196n20
- House Committee on Energy and Commerce and, 46–47
- lobbyists and, 60, 100, 110, 151, 168
- Mineral Policy Act and, 87
- Mining Law and, 87
- moratoria and, 19, 57, 117–119, 142, 178
- National Housing Act and, 108
- National Interstate and Defense Highways Act and, 111–112
- oil embargoes and, 90, 113–115, 123, 125, 143, 183, 198n2, 199n7
- Outer Continental Shelf (OCS) and, 101, 126–127, 131–134, 136, 172–173
- Pact of Achnacarry and, 90
- Parking Standards and, 109–110
- policy recommendations for, 158–170
- Regional Citizens' Advisory Councils and, 166
- Savings and Loans and, 144
- State Lands Act and, 97
- Subcommittee on Oversight and Investigations and, 46–47
- Submerged Lands Act and, 101
- Tidelands Controversy and, 98–101
- U.S. Coast Guard and, 37–39
- U.S. Congress and, 18–19 (*see also* U.S. Congress)
- U.S. Department of the Interior and, 9, 44, 51, 87, 89, 93, 101, 104, 117–130, 138–141, 144, 165, 172
- U.S. Government Accountability Office and, 18–19, 149, 192n10
- U.S. Minerals Management Service and, 16, 48, 50–57, 61, 129–130, 137–141, 165–168, 200nn15,16
- War Revenues Act and, 85

Risers, 29
Risk management, 88, 129, 138
- atrophy of vigilance and, 35–39, 41–42, 159
- BP policies and, 43–51, 55, 60
- cement bond log and, 49
- centralizers and, 48–49
- clean-up and, 158, 162, 165
- cutbacks and, 38
- Emergency Response Team and, 38–39
- *Exxon Valdez* and, 34–35
- House Committee on Energy and Commerce and, 46–47

Lloyd's Register, 46, 52
maximum likelihood estimation and, 35
organizational factors and, 33–39, 41–47, 161–164, 165–170
safety and, 33–51 (*see also* Safety)
skimmers and, 14, 53–54, 154–157
Subcommittee on Oversight and Investigations and, 46–47
technology and, 14–15, 159–161
toxic chemicals and, 41, 168
Rockefeller, John D., Sr.
business acumen of, 75
challenges to supremacy of, 102–103
Cleveland location and, 76–77
competitors of, 77–78, 82, 84
cooperative strategies of, 77–78
moral code of, 75
philanthropic causes of, 75
production phase and, 82
railroads and, 76–80, 83
refineries and, 75–77, 82
Standard Oil Company and, 77–84
Texas and, 79–80
work ethic of, 75
as world's richest man, 75
Rocky (oil platform code name), 9
Rocking horse pumps, 31
Rock oil, 64–66, 176
Roosevelt, Franklin, 99
Roosevelt, Theodore, 79, 86–87
Rothschilds, 77–78
Royalties, 18–19, 52, 56, 59, 121, 148–150, 171, 186
Ruckelshaus, William, 125

Sabotage, 71
Safety, 166
atrophy of vigilance and, 35–39, 41–42, 159
benzene and, 41
BP's casual approach to, 34–61, 161–162
bypassing procedures and, 45, 49
cement bond log and, 49
centralizers and, 48–49
cost-cutting, cutbacks and, 15, 37, 38, 42, 169, 193n12
Emergency Response Team and, 38–39
fail-safe conditions and, 15–16
false alarms and, 45
fines and, 41–42, 109, 162, 170, 193n13
House Committee on Energy and Commerce and, 46–47
maximum likelihood estimation and, 35
OSHA and, 41–42, 167–170, 193n13, 201n13
Process Safety Management for Highly Hazardous Chemicals and, 168
risk management and, 14–15, 33–39, 43–51, 55, 60, 88, 129, 138, 158, 162, 165
shear ram and, 159–160, 164
Subcommittee on Oversight and Investigations and, 46–47
technology and, 15–16
tetraethyl lead and, 88–89

Safety, 166 (cont.)
toxic chemicals and, 41, 168
U.S. Coast Guard and, 37–39
U.S. Minerals Management Service and, 16, 48, 51–52, 55, 61, 129–130, 139, 141, 167–168, 200nn15,16
Safety Award for Excellence (SAFE) award, 52, 165
St. Louis Post-Dispatch, 38
Salt domes, 81, 92, 94
Salt water, 93, 96, 98
Salt wells, 27, 67
Sampson, Anthony, 68
Sandstone, 22
Santa Barbara Morning Press, 124
Santa Barbara oil spill, 10, 91, 123–126, 153–154, 173
Satellite imaging, 12
Saucier, Michael, 55
Saudia Arabia, 103
Schlumberger, 49
Scotland, 90
Sea bottoms, x, 2, 22, 99, 101
Sealing, 29, 32, 43, 48, 54, 63, 159
Sedimentary rock, 21–26
Seeps, 25, 63–67, 116
Seismic technology, 26–27, 94, 148
Semisubmersible rigs, x
Senate Committee on Health, Education, Labor and Pensions, 169
Seneca Indians, 65
Seneca Oil Company, 64–67
Sepulvado, Ronald, 44–45
Sexual promiscuity, 51–57
Shah of Iran, 106
Shales, 22, 24
Shear ram, 159–160, 164
Shell Oil, 90, 103
Sherman Antitrust Act, 78–79, 84, 177
Shipworm (*terredo*), 93
Shivers, Allan, 100
Shortages, 85–86, 88, 115–116, 123, 143
Shrimp and Petroleum Festival, 132
Silent Spill, 13
Silliman, Benjamin, 65
Silt, 21–22, 94–95 135
Skimmers, 14, 53, 154–157
Smith, Billy, 68
Social multiplier effect, 137, 166
Solar power, 5
Solberg, Carl, 84
Source rocks, 23
South Dakota, 25, 94
Soviet Union, 89, 110–111, 114, 150
Speculation, 66, 70, 81
Spindletop, 79, 91, 172–173
blowout of, 80
declining production of, 83
derricks of, 81
environmental issues and, 81, 83
Gulf Oil and, 92
as gusher, 80–83, 116
as largest reservoir discovered at time, 80
limitations of, 83
oil prices after, 81
profit and, 81
size of, 82
Texas Company and, 95
U.S. Market dominance and, 102

Spindletop-Gladys City Boomtown Museum, 83
Standard Oil Company, 98
  Archbold (John) and, 80
  Hogg (James) and, 80, 82
  Interstate Commerce Law and, 78
  Nazi Germany and, 104
  Rockefeller (John) and, 77–84
  Teagle (Walter) and, 90
  tetraethyl lead and, 88–89
  Texas and, 79–80
Standard Oil Trust, 77–79, 84
State Lands Act, 97
Steam power, 36, 72, 76, 93
Stevenson, Adlai, 100
Streetcars, 109
Stripper operations, 83, 195n10
Stupak, Bart, 46
Subcommittee on Oversight and Investigations, 46–47
Submarines, 38, 105
Submerged Lands Act, 101
Submersible rigs, 95–96
Subsidies, 150–151, 182–184
Sunk costs, 95
Sunlight, 21–22
Syria, 103

Taft, William Howard, 79, 86–87
Tariffs, 90
Taxes
  corporate profits and, 18, 83–85
  highways and, 111
  incentives and, 150–152
  injustice to taxpayers and, 18–19, 145–150
  tariffs and, 90
  War Revenues Act and, 85
Teagle, Walter, 90
Teamsters, 71
Teapot Dome, 88, 125
Technology, 49, 92, 115, 131–132, 177
  blow-out preventers, 10–11, 15, 29, 44–45, 55, 93, 159–161, 164
  catalytic converters, 89
  clean-up and, 153–156
  computers, 27
  cost-cutting and, 15, 37, 42, 169, 193n12
  Deepwater Horizon and, ix–x, 15, 19
  Drake engine, 69
  drive pipe, 67
  efficiency and, 5, 28, 73, 84, 88, 143, 151, 181–183
  Exxon Valdez and, 36
  global positioning, x
  management of, 15
  movable, 93–94
  need for new, 183
  offshore drilling methods and, 91–96
  patents and, 65, 87, 95
  Peter Principle and, 19–20
  safety and, 15–16
  satellite imaging, 12
  seismic, 26–27, 94, 148
  semisubersible rigs, x
  skimmers, 14, 53, 154–157
  steam power, 36, 72, 76, 93
  submersible rigs, 95–96
  tetraethyl lead, 88–89
  thrusters, x, 45
  whale oil and, 63–66, 71
Tethers, x
Tetraethyl lead, 88–89
Texas, 117, 150, 152, 169

Texas (cont.)
  Barton (Joe) and, 19, 58, 60, 194n41
  Bush (George W.) and, 56
  Hogg (James) and, 80, 82
  inland sea of, 25
  judicial conflicts of interest and, 59
  marginal sea and, 99
  oil prices and, 90
  production restriction and, 89
  Rockefeller (John) and, 102
  Spindletop and, 79–83, 91–92, 95, 102, 116, 172–173
  Submerged Lands Act and, 101
  tax revenue of, 18
  Tidelands Controversy and, 98–101
  wildcatters and, 4, 82–83, 152
Texas City refinery explosion, 41, 44, 45, 169
Texas Company (Texaco), 82, 95, 177
Thomas, Henry, 116, 122
Thrusters, x, 45
Tidal exchange, 97
Tidelands Controversy, 98–101
Titanic (ship), 15, 36
Torrey Canyon spill, 156
Toxic chemicals, 41, 168
Train, Russell, 119
Trans-Alaskan Pipeline Authorization Act, 123
Trans Alaska Pipeline System (TAPS), 3, 102, 118, 120, 122–123
Transocean, 33, 53
  casual approach to safety by, 34, 45–46
  Crawford (Truitt) and, 43
  exclusion and, 165
  Feldman (Martin) and, 59
  Hay (Mark) and, 45
  lawmakers' holdings in, 58–59
  Moody's rating of, 46
  Safety Award for Excellence (SAFE) award and, 52
  shear ram and, 159–160
  U.S. Minerals Management Service (MMS) and, 52
  Williams (Mike) and, 45
Transparency, 162–163
Transportation, 63, 72, 75, 121, 183, 196n14
  Alaska and, 118
  Chicago and, 76
  Interstate Commerce Law and, 78
  modernizing of, 109
  Panama Canal and, 86
  public transit and, 109–110
  railroads and, 71, 76–80, 83, 86, 95, 109
  roads and, 95 (*see also* Automobiles)
  Rockefeller (John) and, 76–80, 83
  streetcars and, 109
Tropical Storm Isidore, 187
Truman, Harry, 99
Tunisia, 18
Turkish Petroleum Company, 103

Udall, Stewart, 117–119
Union Oil Company of California, 118, 177
United States
  American Way of Life and, 185–189
  automobile popularity in, 107

Bush (George H.W.) and, 6, 141–142
Bush (George W.) and, 2–3, 6, 53, 56, 142, 178
Carter (Jimmie) and, 5, 125–127, 140, 143
Civil War and, 75–76
consumption levels of, 84–85
decreased production levels of, 179
as dominant-oil producing nation, 1–2
drilling rights and, 18
early use of oil and, 63–66
Eisenhower (Dwight) and, 1, 100–101, 106, 110
energy crisis of, 113, 122–123
energy independence and, 4–6, 20, 85–86, 113–128, 144, 180, 183
Exclusive Economic Zones and, 2
Federal-Aid Highway Acts and, 110–111
Federal Trade Commission and, 90, 196n20
Ford (Gerald) and, 5, 125–126
foreign oil and, 5–6, 90, 113–114, 121–122
Franklin (Benjamin) and, 183
future and, 178–184
Gilded Age and, 76
Great Plains and, 24–25
Harding (Warren) and, 87, 89
Harrison (Benjamin) and, 78
Hoover (Herbert) and, 89
housing and, 108–110
Interstate Commerce Law and, 78
invasion of Afghanistan by, 1
Iraq and, 1–2, 4, 66, 71, 101–103, 114
judicial branch and, 59
loss of petroleum preeminence of, 101–112
McKinley (William) and, 78
marginal sea and, 99
National Housing Act and, 108
National Interstate and Defense Highways Act and, 111–112
natural resources of, 63–66, 71
Nixon (Richard) and, 5, 113, 115, 119, 122–123, 125, 172
Obama (Barack) and, xii, 6, 19, 53, 56–59, 147, 150, 178
oil embargoes and, 90, 113–115, 123, 125, 143, 183, 198n2, 199n7
oil revenue rates of, 18–19
Outer Continental Shelf Deep Water Royalty Relief Act and, 19, 149
ownership of oil from Macondo project, 18
Pact of Achnacarry and, 90
Project Independence and, 113, 115, 123
railroads and, 71, 76–80, 83, 86, 109
Reagan (Ronald) and, 5–6, 51, 127–128, 136, 138–139, 143–144
Roosevelt (Franklin) and, 99
Roosevelt (Theodore) and, 79
Stevenson (Adlai) and, 100
Taft (William) and, 79, 86–87
Truman (Harry) and, 99
whale oil and, 63–66, 71
Wilson (Woodrow) and, 84

Unocal (Union Oil of California), 177
U.S. Coast Guard, xii
  atrophy of vigilance and, 37–39
  Deepwater Horizon spill and, 10, 43–44, 50
  safety issues and, 37–39
  U.S. Minerals Management Service (MMS) and, 55, 167–168
U.S. Congress
  Alaska and, 117, 120–122, 126
  Barton (Joe) apology and, 19, 58, 60, 194n41
  BP's casual approach to safety and, 41, 43–44, 47, 50
  clean-up and, 169–170
  Congressional Budget Office and, 150–151
  Eisenhower (Dwight) and, 100
  Federal-Aid Highway Acts and, 110–111
  House Committee on Energy and Commerce and, 46–47
  Interstate Commerce Law and, 78
  Mineral Policy Act and, 87
  National Interstate and Defense Highways Act and, 112
  oil leases and, 89, 141 (*see also* Leases)
  Outer Continental Shelf Deep Water Royalty Relief Act and, 19, 149
  Outer Continental Shelf Lands Act (OCSLA) and, 101
  Outer Continental Shelf Lands Act Amendments (OCSLAA) and, 127
  policy recommendations for, 173, 178, 185
  poor responses of, 57, 169
  railroads and, 76
  Senate Committee on Health, Education, Labor and Pensions, 169
  Standard Oil Trust and, 84
  Submerged Lands Act and, 101
  Stupak (Bart) and, 46
  Subcommittee on Oversight and Investigations and, 46–47
  subsidies and, 150–151
  U.S. revenues and, 18
  Waxman (Henry) and, 46
U.S. Department of the Interior, 4, 44, 101, 165
  Alaskan oil and, 117–118, 120–121, 123, 125–128
  area-wide leasing and, 172–173
  Bureau of Land Management and, 117–118, 138, 172
  Fall (Albert) and, 87
  General Land Office and, 92–93
  Ickes (Harold) and, 104
  leases and, 130 (*see also* Leases)
  Mississippi Canyon and, 9
  Naval Reserves and, 87
  offshore wells and, 138–141, 144
  Rankin (John) and, 172
  Reagan and, 138–139
  sanity issues and, 139, 199n6
  Udall (Stewart) and, 117–119
  Watt (James) and, 51
U.S. Department of Defense, 122
U.S. Fuel Administration, 84–85

U.S. Geological Survey, 12, 85, 138, 172
U.S. Government Accountability Office, 18–19, 149, 192n10
U.S. Minerals Management Service (MMS), 16, 48, 50, 166, 200nn15,16
Alaska and, 138
area-wide leasing and, 139–140, 144–149, 172–173, 186
BP and, 51–57
Bureau of Ocean Energy Management, Regulation and Enforcement and, 57, 167–168
Bush (George H.W.) and, 141–142
Bush (George W.) and, 56
California and, 130–132
Cheney (Richard) and, 56
corruption and conflicts of interest in, 51–57, 61
Energy Plan and, 56
false claimsby, 55
Florida and, 130–131
Louisiana and, 130, 131–137, 140, 142–143
party-line views within, 142
offshore wells and, 129–137, 140–143
public education programs and, 137–138
reform and, 167–168
risk perceptions and, 129–130
Safety Award for Excellence (SAFE) award and, 52, 165
safety records of, 52–53, 165–168
sex/drug issues at, 51–57
spiral of stereotypes and, 140
as toothless watchdog, 51
U.S. Coast Guard and, 55
Watt (James) and, 51, 129, 138–139
U.S. Navy, 86
U.S. Supreme Court, 79, 89, 99–101
U.S. Treasury Department, 150–151, 171, 173

Vanderbilt family, 76
Vanity Fair, 40
Venezuela, 89
Lake Maracaibo field and, 93–95
offshore wells and, 91–95
OPEC and, 114
Verlger, Philip, 52
Vermilion Bay, 94
Vietnam, 18
Viscosity, 23, 31
Voids, 22

Wall Street, 40, 46
Wall Street Journal, 20
War Revenues Act, 85
Washington, State of, 140
Washington Post, 14, 40, 42–43, 46, 55–57, 59
Waste, 3–4, 38, 83
Water cannons, xii
Watergate scandal, 125, 198n15
Water Pollution Control Act, 41
Watt, James
area-wide leasing and, 139–140, 144–149, 172, 186
environmental issues and, 51, 54, 128–129, 137–138, 172–173
Waxman, Henry, 46
Wetlands, 12
bayous, 94, 97, 134–135
marshes, 94, 96–98, 133, 156

Whale oil, 63–66, 71
Wildcatters, 4, 82–83, 152
Williams, Mike, 45
Williston basin, 25
Wilson, Woodrow, 84
World Bank, 32
World War I, 84, 105
World War II, 99, 116, 176
  oil as reason for, 104–105, 109, 111
  Pearl Harbor and, 104
Wyoming, 18, 86, 88, 150

Yellow Coach, 109
Yukon River, 123